Variations in
human physiology

Variations in human physiology

Edited by
R.M. Case

Contributions by
D.E. Evans, H.O. Garland, R. Green,
R.A. Little, D.S. Minors, J.M. Waterhouse

*The editor and contributors are all members of
the Department of Physiology, University of Manchester,
except Dr Little, who is a member of
the MRC Trauma Unit, University of Manchester,
and Honorary Lecturer in Physiology*

Manchester University Press

Copyright © Manchester University Press, 1985

First published in 1985
by Manchester University Press
Oxford Road, Manchester M13 9PL, UK
and 51 Washington Street, Dover
New Hampshire 03820, USA

Reprinted 1986

British Library cataloguing in publication data
Variations in human physiology. — (Integrative
 studies in human physiology)
 1. Human physiology
 I. Case, R.M. II. Series
 612   QP34.5

Library of Congress cataloging in publication data
Main entry under title:
Variations in human physiology.
 Includes bibliographies and index.
 1. Adaptation (Physiology) 2. Human physiology.
3. Variation (Biology) I. Case, R. Maynard (Richard
Maynard) II. Evans, D. E. [DNLM: 1. Adaptation,
Physiological. QT 140 V299]
QP82.V37 1985   612   85-11301

ISBN 0-7190-1086-1
ISBN 0-7190-1732-7 (pbk.)

Typeset in Hong Kong
by Graphicraft Typesetters Ltd.
Printed in Great Britain by
Butler & Tanner Ltd, Frome and London

Contents

Preface

Physiology courses at Manchester University, like many more elsewhere, first cover the main areas (such as body fluids, excitable tissues, the cardiovascular system, etc.) in a systematic way.

In such courses, attention is focused on a standard human subject who lives at sea level, is forever healthy, 70 kg in weight and 18–21 years old. The subject's various systems are examined as separate entities, each seen as a model. Here — a nervous system; there — a kidney; occasionally a piece is borrowed from one model to complete another, but is hastily returned. Popular texts cover this material more or less successfully.

'Yet one man in his time plays many parts ... hurt with weapons ... subject to diseases ... healed ... warmed and cooled by winter and summer' (*sic*, with apologies). All are born; some give birth. Some succumb to a hostile environment; the rest survive to be old. Hence, towards the end of a course, a little integration is attempted. Now the student is at a loss. Smaller texts rarely have space for more than systems, the essential foundations. Larger and more specialised books are in short supply, and are sometimes intimidating.

This book is an attempt to meet this need. We presume a basic knowledge of 'standard' physiology such as one acquires in a preclinical course in medicine or dentistry, or during the early years of a science course in physiology. Where possible, human physiology is described except when comparative studies are particularly helpful or human data are not available. Each chapter can be read in isolation, though cross-references between chapters emphasize the integrated nature of physiology. A short bibliography accompanies each chapter to guide the interested reader towards a more detailed study of a given topic.

One difficulty in writing this book has been to find a suitable title! *The Physiology of Altered Circumstances* (our first attempt) contains too many etymological inexactitudes. *Apply Your Physiology* and *Not the 70 kg Man* might have been misunderstood. The chosen title reflects the main message of this book, namely that much variation exists in the physiology of normal healthy people: indeed it is essential for their survival.

In conclusion, we apologise (to teachers) for encroaching upon material much beloved by examiners. They, like us, continually require students to apply their knowledge of physiology to the problems imposed by 'unusual' environments, or to unravel the complexities of exercise: we shall now have to discover, or invent, new environments and activities.

Acknowledgements

For their comments on the manuscript, we thank each of the following: Damian Longson, David Throssell and Catherine Pickup, undergraduate students of Physiology at Manchester University, who read the whole book; Professor Ian Houston, Department of Child Health (Chapter 2); and Dr Iouan Davies, Department of Geriatric Medicine (Chapter 3).

We thank Richard Neave and his staff in the Department of Medical Illustration for drawing many of the diagrams, and Air Commodore P. Howard for supplying Figure 7.1.

We also thank authors and publishers for granting us permission to reproduce copyright material: the source of such material is given in the legends of those figures that have been copied.

Finally we thank Mrs Sheilah Long and Miss Anne Kaye for their skill and patience in typing the manuscript, and the staff of our own University Press for their efficient processing of the text.

Chapter 1

Maternal adjustments to pregnancy

Summary

Human physiology is altered in pregnancy, primarily to cope with the demands of
the developing fetus. This chapter describes the changes that occur in the major
physiological systems of the mother during gestation, beginning with a brief outline
of the major endocrine changes of pregnancy itself. Such changes provide a background against which the alterations in maternal physiology must be considered.

Major changes in the circulatory system during gestation include an increased
plasma volume with a smaller increase in red cell mass, producing the characteristic
'physiological anaemia' of pregnancy. Cardiac output is also increased to allow an
augmented regional blood flow to many maternal organ systems. Total peripheral
resistance falls to maintain blood pressure. The enhanced regional blood flow has
many consequences. The increased renal blood flow, for example, raises the glomerular filtration rate in the kidney, and an enhanced tubular reabsorption is required
to prevent large salt and water losses. Nevertheless, glycosuria and amino aciduria
are both characteristics of normal pregnancy.

Changes in maternal respiration are necessary to cope with the increasing oxygen
consumption of pregnancy. Pulmonary ventilation is thus enhanced, and this also
serves to facilitate the removal of CO_2 from the developing fetus. The increased
nutritional requirements of pregnancy are to some extent satisfied by a stimulation
of appetite, together with an increased intestinal absorption of certain substances.
Major changes in gastrointestinal motility, however, are due to increased progesterone concentrations.

The final two sections of this chapter consider changes in body weight and
metabolism during pregnancy. Body weight changes are analysed in terms of maternal
components and the products of conception. The former comprise increases in
uterine and breast tissue, extracellular fluid and fat. The latter is made up of the
fetus itself, together with the placenta and amniotic fluid. There also occur changes
in many aspects of metabolism, which both provide for the developing fetus and at
the same time protect the pregnant mother against any major nutritional deficiencies
in her diet.

1.1 Introduction

Human pregnancy is accompanied by dramatic changes in many of the
physiological systems of the mother. Many of these changes can be directly
related to the requirements of the fetus. All occur against a background of
the characteristic endocrine changes of pregnancy itself. Although it is not
possible to explain all of the maternal changes by the underlying endocrinology, it is appropriate to begin this chapter with a brief summary of the
major hormonal changes that occur in human gestation.

Figure 1.1 shows the changing endocrine profile for oestrogen and

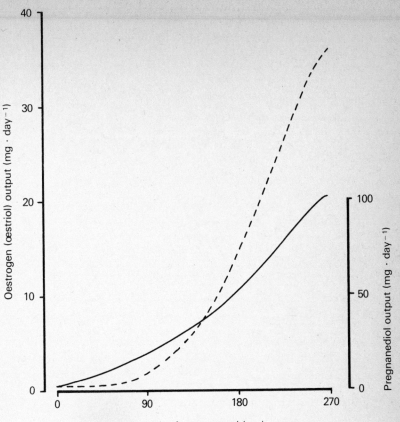

Fig. 1.1   Oestrogen (----) and progesterone (——) concentrations during human pregnancy. Daily excretion patterns of the oestrogen, oestriol and the urinary metabolite of progesterone, pregnanediol.

progesterone during human pregnancy. Both steroids are secreted initially by the corpus luteum, but after the sixth week of gestation, the placenta assumes the major role in their production. The placenta is also the source of two protein hormones, human chorionic gonadotrophin (hCG) and human placental lactogen. Section 2.2.4 describes their changing circulating levels throughout pregnancy.

The maternal pituitary increases in size by 30–50 per cent in pregnancy, and its secretion of prolactin, adrenocorticotrophic hormone (ACTH) and melanocyte-stimulating hormone (MSH) are all enhanced. Growth hormone and gonadotrophin release, however, are both inhibited, the former by placental lactogen, the latter by placental steroids. The increase in ACTH during gestation is accompanied by a rise in plasma cortisol.

Aldosterone secretion by the adrenal cortex is also enhanced, both by the salt-losing effect of progesterone on the kidney, and through stimulation of the renin–angiotensin system. Circulating parathormone (PTH) concentrations rise early in pregnancy or around mid-term.

1.2 Circulatory changes

1.2.1 *Plasma volume*
Plasma volume normally increases during human pregnancy to a value around 50 per cent above the non-pregnant level by the start of the third trimester (Figure 1.2). The precise increase is related to fetal mass, and may approach 90 per cent in women with triplets or quadruplets. It is not known whether this increase is the direct result of an enhanced fluid intake, reduced output, or shift between body fluid compartments. Certainly there is a rise in sodium-retaining steroids (see 1.1), and the kidney reabsorbs more fluid (see 1.3.2), but glomerular filtration rate (GFR) is also increased. The increased plasma volume of pregnancy may contribute to the altered GFR (see 1.3.1).

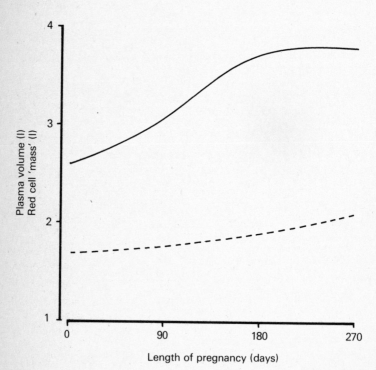

Fig. 1.2 Changes in plasma volume (——) and red cell mass (– – – –) during human pregnancy.

1.2.2 *Blood cells*

The total volume of red cells in the circulation (red cell mass) normally increases by about 20 per cent during pregnancy (Figure 1.2). As with plasma volume, this increase varies with the size of the fetus and may be three times this value in women with multiple births. As Figure 1.2 shows, the enhanced plasma volume of pregnancy is proportionally greater than the increase in red cell mass. Consequently, the red cell count, haematocrit and haemoglobin concentration are all reduced (Table 1.1). Such changes constitute the 'physiological anaemia of pregnancy'. Other haematological indices (mean cell volume, cell haemoglobin content and cell haemoglobin concentration) do not alter significantly.

Table 1.1. Haematological indices: typical values for pregnant and non-pregnant women.

	Non-pregnant	*Term pregnant*
Red cell count (cells $\cdot$ litre^{-1})	4.7×10^{12}	3.8×10^{12}
Haematocrit (%)	42	33
Haemoglobin concentration (g $\cdot$ dl^{-1})	14	11.5

The increased red cell mass in pregnancy is caused by an enhanced red cell production. Plasma erythropoietin concentration is raised, and the action of this hormone may be augmented by human placental lactogen. Iron and folic acid requirements increase during pregnancy to cope with the enhanced erythropoiesis (see Table 1.2 and Section 1.5.1.2).

The total white cell count rises from a normal value of 9×10^9 to 11×10^9 cells $\cdot$ litre^{-1} during the third trimester of human pregnancy. This is due almost exclusively to an increased production of neutrophils. These cells constitute the largest fraction of the total leukocyte population and their formation is thought to be stimulated by rising oestrogen levels. Other white cells remain steady or fall as gestation proceeds. Elevated hCG, prolactin and oestrogens may suppress lymphocyte function and cellular immunity. This could help prevent fetal rejection but will, at the same time, render the mother more susceptible to infection. Indeed, there is an increased incidence of influenza, rubella and hepatitis in pregnancy.

The platelet count falls slightly during gestation. Several clotting factors, however, are increased, possibly as a result of the increased steroid concentrations. These include fibrinogen, and factors VII, VIII, IX, X and XII. Stimulation of the clotting mechanism helps to prevent excessive haemorrhage during separation of the placenta from the uterus at term.

1.2.3. *Cardiovascular system*

1.2.3.1 *Cardiac output, blood pressure and total peripheral resistance.*
Cardiac output increases about 40 per cent above its non-pregnant value by
the end of the first trimester and remains elevated until term. The increase
in cardiac output is a consequence of an increase in both components: heart
rate, typically from 70 to 85 beats $\cdot$ min^{-1}, and stroke volume, from 60 to
70 ml.

Arterial blood pressure, in contrast, is relatively unaffected. Systolic
pressure shows little change during pregnancy, although diastolic pressure
falls by about 10 mmHg (1.3 kPa) so pulse pressure is increased. In order to
keep blood pressure constant, total peripheral resistance decreases to compensate for the raised cardiac output. This results primarily from peripheral
vasodilatation, although arteriovenous anastomoses in the uterus may also
open. Several mechanisms have been suggested to account for the vasodilatation in pregnancy. Oestrogens and progesterone may inhibit arteriolar
smooth muscle tone; the effects of pressor agents such as angiotensin may
be reduced; or active vasodilator substances may be released.

Venous blood pressure in the lower limbs increases three- to fourfold in
pregnancy as a result of compression of the vena cava by the distended
uterus. The elevated venous pressure often causes lower limb oedema, and
leg veins may become varicose.

1.2.3.2 *Regional blood flow.* The distribution of the increased cardiac output of pregnancy is shown in Figure 1.3. The pregnant uterus is a major site

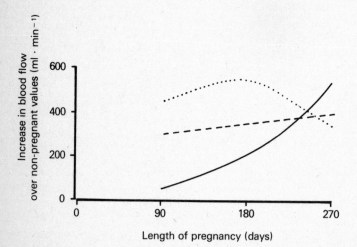

Fig. 1.3   Distribution of increased cardiac output during pregnancy. Symbols:
——, uterus; – – –, skin; . . . ., kidney. Adapted from Hytten, F. & Chamberlain, G.
(1980). *Clinical Physiology in Obstetrics.* Blackwell: Oxford.

for the increased circulation, and uterine blood flow increases progressively throughout gestation, matching the requirements of the growing fetus. A value ten times the non-pregnant level is achieved by term. Skin blood flow also shows a progressive increase during pregnancy. This will serve to dissipate heat from both the fetus and the mother, whose basal metabolic rate is raised in pregnancy.

Altered renal blood flow, however, shows a different pattern. A rise of up to 80 per cent above non-pregnant values occurs during the first trimester, and this is maintained, or declines towards term (see also 1.3.1). Blood flow to the breasts and gastrointestinal tract are both enhanced in pregnancy.

1.3 Renal changes

1.3.1 *Glomerular filtration rate and renal blood flow*
Both glomerular filtration rate (GFR) and renal blood flow are increased by 30–80 per cent in human pregnancy, although their precise pattern of

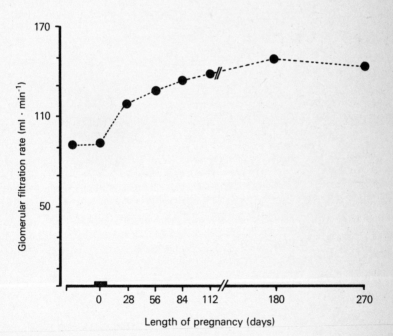

Fig. 1.4 Changes in glomerular filtration rate in women prior to conception and during pregnancy. The bar on the horizontal axis represents the last menstrual period. Adapted from Davison, J.M. (1978). *Yale Journal of Biology and Medicine*, **51**, 347–9 and Davison, J.M. & Hytten, F. (1974). *Journal of Obstetrics and Gynaecology of the British Commonwealth.* **81**, 588–95.

change remains controversial. Recent studies indicate that changes in GFR are established at a much earlier stage of pregnancy than was previously thought, maximum values being attained 6–8 weeks after conception (Figure 1.4). Interestingly, such changes may not be apparent in women destined for early miscarriage. However, in normal pregnancies, the increases are maintained until term. Renal plasma flow follows a similar pattern to GFR, although some studies have shown it to decline slightly in the third trimester.

The mechanism responsible for the elevated GFR of human pregnancy is not fully understood. However, animal studies suggest it is exclusively the result of the enhanced renal plasma flow rate, since the other determinants of glomerular filtration are unaffected. The increased renal plasma flow may in turn be at least partially dependent upon the expanded plasma volume of pregnancy (see 1.2.1). Some of the endocrine changes described earlier (see 1.1) may be responsible, either directly or indirectly, for the renal haemodynamic changes of pregnancy. Adrenal steroids and oestrogens, for example, cause sodium (and fluid) retention in man; prolactin elevates GFR in human and animal studies.

1.3.2 *Renal tubular function*
1.3.2.1. *Salt and water reabsorption.* The changes in GFR during human pregnancy described above will result in an enhanced filtered load of water and solutes. Assuming a normal plasma sodium concentration of 140 mmol·litre^{-1}, a 50 per cent increase in GFR from 180 to 270 litres per day will result in an extra 12.6 mol of sodium being filtered daily. In order to prevent the loss of such large quantities of sodium, the renal tubular reabsorption of salt and water is also increased. Indeed, the enhanced renal sodium reabsorption, when coupled to an increased intake, produces a net retention of almost 1 mol of sodium during human pregnancy. Around half of the retained sodium is used by the fetus.

A stimulation of the renin–angiotensin–aldosterone system may contribute towards the increase in renal sodium handling during pregnancy. It will also offset the potential natriuretic effect of high progesterone concentration. In addition, recent animal studies have shown the proximal segment of the renal tubule to elongate in pregnancy, thus increasing the tubular area available for reabsorption. Comparable studies are not possible in humans, of course, although excretory urograms have indicated an increased overall renal size in pregnancy.

The control of anti-diuretic hormone (ADH) secretion is altered in pregnancy. In women, plasma osmolality falls by about 10 mosmol·kgH$_2$O^{-1} during the first trimester. Normally, such a change would be accompanied by a reduced ADH secretion, and a subsequent diuresis would restore plasma osmolality to normal. This does not happen in pregnancy. It has therefore been suggested that the osmoreceptors responsible for controlling ADH release are 'reset' at a lower value during gestation. Indeed, the ability to excrete a water load declines during the third trimester. This altered renal response, coupled to an enhanced fluid intake, accounts for the increase in total body water during pregnancy. Gains of 6–8 litres have been reported, around half of which is accounted for in the products of conception (fetus, placenta and amniotic fluid).

1.3.2.2 *Handling of glucose, amino acids and vitamins.* Glycosuria is a characteristic of normal human pregnancy, and urinary losses of glucose may increase tenfold by the end of the first trimester. The excretion of other sugars (fructose, lactose, ribose, xylose) is also enhanced in pregnancy.

Until recently it was generally assumed that all glucose reabsorption in the kidney occurs in the proximal segment of the renal tubule. Consequently, the glycosuria of pregnancy has been attributed either to the increased filtered load of glucose exceeding the proximal tubule reabsorptive capacity, or to a decrease in the reabsorptive capacity itself. However, recent animal studies have shown that more distal segments of the nephron normally reabsorb small amounts of glucose, and that it is this reabsorption that is impaired in pregnancy. Whatever the sites for altered glucose handling in human pregnancy, the controlling mechanisms remain unknown.

As well as losing more sugars in her urine, the pregnant woman also excretes larger quantities of most amino acids. Arginine, asparagine and isoleucine are the main exceptions to this rule. Overall, urinary amino acid losses in late pregnancy may approach $2 \text{ g} \cdot \text{day}^{-1}$. This is normally compensated by an increased dietary protein intake during gestation (see 1.5.1.2). The reason for the increased renal loss is not known, although tubular reabsorption is thought to be depressed.

A similar explanation may account for the increased excretion of water-soluble vitamins (ascorbic acid, folate, nicotinic acid) and uric acid during pregnancy.

1.4 Respiratory changes

1.4.1 *Oxygen consumption*
Oxygen consumption increases by around 15 per cent during pregnancy. The components of this increase are shown in Figure 1.5. The increased demands of the fetus and placenta are progressive throughout pregnancy, whereas altered maternal requirements are established by the end of the first trimester. The increased oxygen consumption can be related to the enhanced oxygen-carrying capacity of the blood since both red cell mass and plasma volume increase during gestation (see 1.2.1 and 1.2.2).

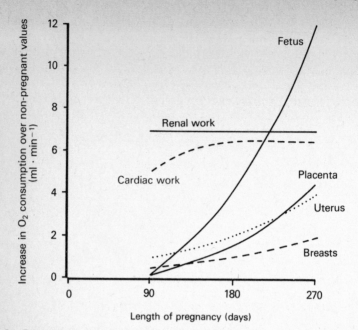

Fig. 1.5   Components of the increased oxygen consumption of pregnancy. Adapted from Hytten, F. & Chamberlain, G. (1980). *Clinical Physiology in Obstetrics.* Blackwell: Oxford.

Increased oxygen requirements in pregnancy have obvious respiratory implications, which are considered in the following sections.

1.4.2 *Pulmonary ventilation*

Pulmonary ventilation increases by about 40 per cent during pregnancy to around 10 litres $\cdot$ min^{-1}. This increase considerably exceeds the increase in oxygen consumption described above. It is due exclusively to an increase in tidal volume; respiratory rate is unaltered.

Figure 1.6 shows the altered tidal volume together with other changes in lung volume parameters that occur in pregnancy. The 40 per cent increase in tidal volume contributes to an overall increase in inspiratory capacity of around 300 ml. Vital capacity, however, shows a much smaller increase, as the expiratory reserve volume falls. Since the residual volume is also reduced, the functional residual capacity decreases by around 20 per cent of its non-pregnant value. The inspiratory capacity has, therefore, increased at the expense of the expiratory reserve, so that the lung is relatively more collapsed at the end of a normal expiration during pregnancy.

The lung volume changes described above are accompanied by character-istic alterations in the shape of the thorax. The diaphragm is pushed

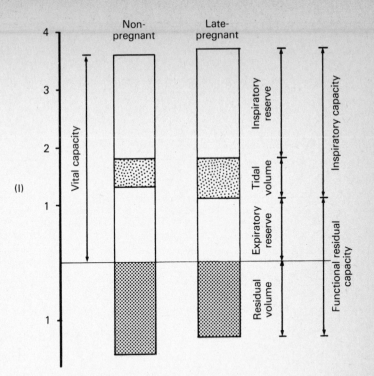

Fig. 1.6 Lung volumes and their alterations in pregnancy. From Hytten, F. & Chamberlain, G. (1980). *Clinical Physiology in Obstetrics*. Blackwell: Oxford.

upwards by about 4 cm and the thoracic cavity increases in diameter by about 2 cm.

1.4.3 *Gas exchange and transport in blood*
An estimate of alveolar ventilation can be made from the data in Figure 1.6. If the tidal volume increases in pregnancy from 500 to 700 ml, and respiratory rate and physiological deadspace remain constant at $15 \cdot min^{-1}$ and 150 ml respectively, then alveolar ventilation increases from 5.3 to 8.3 litres $\cdot min^{-1}$, an increase of 57 per cent. In fact, physiological deadspace is probably also raised in pregnancy, due to dilatation of the smaller bronchioles, and so the figure of 57 per cent is a slight overestimate.

The transfer of gases across the alveolar membrane is retarded during pregnancy. This is thought to be due to the high concentration of circulating oestrogens which structurally alter the alveolar membrane. However, any effect on oxygen transport is offset by the higher alveolar partial pressure of oxygen produced by the increased ventilation. As a result, arterial PO_2 is increased in pregnancy whereas PCO_2 falls, typically from 40 to 30 mmHg

(5 to 4 kPa). The fall in PCO_2 is accompanied by a fall in plasma bicarbonate concentration, and so arterial pH remains at 7.4.

1.4.4 *Control of respiration*

The characteristic hyperventilation of pregnancy described above is due to an increased sensitivity of the respiratory centre to PCO_2; pregnant women respond to a raised PCO_2 by increasing their pulmonary ventilation to a greater extent than do non-pregnant women. This effect is mediated by the increased circulating progesterone concentration. It is an adaptive response inasmuch as it enhances the removal of carbon dioxide from the developing fetus as well as satisfying the increased oxygen requirements of both mother and fetus.

1.5 Gastrointestinal changes

1.5.1 *Dietary intake*

1.5.1.1 *Appetite and thirst*. Increased appetite is a common characteristic of pregnancy and, if food is freely available, the daily calorific intake usually increases by about 10 per cent during gestation. The stimulation of appetite may be due to a central action of progesterone. Animal studies have shown the steroid to have a stimulatory effect on food intake via the feeding centres in the hypothalamus. A reduced plasma glucose concentration in early pregnancy (see 1.7.1) may also stimulate appetite.

In addition to enhanced appetite, pregnant women often report an increased thirst. The cause of this is not known, although angiotensin and prolactin, two hormones with a known dipsogenic action in animals, are both raised in human pregnancy. Any enhanced fluid intake may contribute towards the characteristic expansion of plasma volume in pregnancy (see 1.2.1).

1.5.1.2 *Recommended nutritional allowances*. Table 1.2 presents some recommended nutritional allowances for pregnancy. The increased allowances may be destined for either maternal or fetal usage. For example, iron and folic acid may facilitate maternal erythropoiesis (see 1.2.2). Extra calcium and phosphate are used primarily by the fetus for skeletal development (see 1.7.2). Increased dietary protein will be used for both maternal and fetal growth (see 1.6).

Although such tables have obvious relevance to affluent Western societies, many pregnancies in Third World countries proceed despite chronic maternal malnutrition. The developing fetus is thus believed to be much less vulnerable to its mother's dietary inadequacies than was once believed. Maternal metabolic changes during pregnancy (see 1.7) may compensate for many dietary shortcomings.

Table 1.2. Recommended daily nutritional allowances for women (compiled from data produced by the Department of Health and Social Security (UK) and the National Academy of Sciences National Research Council (USA)).

	Non-pregnant	Pregnant
Energy (kJ)	8820	9660
Protein (g)	55	63
Calcium (g)	0.7	1.2
Phosphorus (g)	0.8	1.2
Iron (mg)	12	15
Vitamin D (IU)	100	400
Ascorbic acid (mg)	40	60
Folate (μg)	400	800
Nicotinic acid (mg eq)	14	17
Riboflavin (mg)	1.4	1.7
Thiamine (mg)	1.0	1.1
Vitamin B_{12} (μg)	5	8

1.5.1.3 *Pica and aversion.* Pica, an abnormal craving for unusual foods, is experienced by many pregnant women. The more bizarre cravings are often for strongly flavoured or highly textured substances, such as pickled onions, coal, chalk and dog biscuits. Unorthodox food combinations may also be favoured. The desire for strongly flavoured food may be related to a dulling of the tastebuds in pregnancy.

A distaste for certain foods may also be apparent in pregnancy. Common aversions include tea, coffee, cigarettes and alcohol. This particular choice may suggest a psychological rather than physiological basis for the aversion, since increasing publicity is now given to the potential adverse effects of drugs on the developing fetus.

1.5.2 *Gastrointestinal organs*

1.5.2.1 *Mouth and oesophagus.* There are few clearly established changes in salivary secretion during human pregnancy. However, the incidence of new caries increases in the third trimester, and this may be due in part to a reduced salivary pH.

In the oesophagus, a reduced competence of the lower oesophageal sphincter may be responsible for an increased regurgitation of gastric acid from the stomach during pregnancy. This could explain the high incidence of heartburn in pregnant women.

1.5.2.2 *Stomach.* Gastric acid secretion is reduced during early and mid-pregnancy in women, and this may explain the reduced incidence of gastric ulcers during pregnancy. Such findings contrast with the situation in laboratory animals, such as rats and mice, where pregnancy is associated with a progressively increased secretion of gastric juice. Elevated oestrogen and progesterone concentrations in human pregnancy may enhance mucus secretion by the neck cells. This will further protect against potential ulceration.

In pregnant women, gastric tone and motility are both reduced. Gastric emptying is thus slowed, and the amount of food remaining in the stomach after a given time increases. Early animal studies suggested a similar picture in pregnant rodents, although this has recently been questioned. The reduction in gastric tone, together with intestinal and colonic atony in pregnant women (see below), is usually attributed to the relaxing effect of progesterone on smooth muscle. More recently, however, the possible involvement of gut hormones has been investigated (see below). Decreased gastric tone and motility may be responsible for the common complaint of nausea in pregnancy. In man, experimentally induced nausea is associated with interruption of normal gastric contractions and reduced stomach tone.

1.5.2.3 *Intestine.* In small mammals, the most conspicuous change in the whole of the gastrointestinal tract during pregnancy and lactation is an enlargement of the intestine. This is the result of both hypertrophy and hyperplasia, and produces an increased area for absorption. At the same time, intestinal motility is reduced.

There is no evidence in pregnant women for a similar intestinal hypertrophy, although the uptake of substances such as calcium and iron may be enhanced. Intestinal propulsion, however, is reduced in women during pregnancy, and the transit time for food is consequently increased. Colonic atony is a contributory factor towards constipation in pregnancy, as is an enhanced fluid absorption in the colon. Both the renin–angiotensin system and prolactin have been implicated in the latter effect.

Although progesterone is commonly held responsible for the reduced intestinal tone in pregnancy, the potential contribution of several gut hormones has recently been considered. Motilin, for example, an intestinal peptide which normally stimulates gastrointestinal smooth muscle, is reduced in human pregnancy. Enteroglucagon, however, which normally slows intestinal transit, is raised. This hormone may also contribute to the gastrointestinal enlargement seen in pregnant rats and mice, since it is capable of stimulating intestinal mucosal hyperplasia. Other potential candidates for the stimulation of intestinal growth in these species include gastrin and prolactin.

1.5.2.4 *Liver.* A progressive hepatic growth is a characteristic of pregnancy in experimental animals. Indeed, the mass of the liver of the term-pregnant mouse is over 75 per cent greater than that of a comparable virgin. Growth is due to both hypertrophy and hyperplasia. Hepatic blood flow also increases. Far less is known about any changes in liver function during human pregnancy. However, the growth and increased blood flow seen in small mammals is not reported in pregnant women. Nevertheless, the general activity of the liver, at the centre of numerous metabolic processes, is much increased.

1.6 Growth and body weight changes

1.6.1 *Total weight gain*
The average total weight gain in normal human pregnancy is around 12.5 kg, although the reported range is wide (-11 to $+23$ kg). The absolute value depends upon a number of factors, including maternal diet, age and non-pregnant body weight. Young mothers, for example, tend to gain more weight than older mothers, heavy women put on less weight than light ones. Figure 1.7 describes the 'standard' rate of weight gain throughout pregnancy, and also shows the components of the total weight gain at term. This breakdown is discussed further below. Of the 12.5 kg gain in body weight, around 60 per cent is fluid, although this value can exceed 70 per cent in women with oedema.

1.6.2 *Maternal component*
The maternal component of the total weight gain at term (as shown in Figure 1.7) averages 7.7 kg, and comprises increases in uterine and breast tissue, extracellular fluid and fat. The uterus and breasts increase progres-

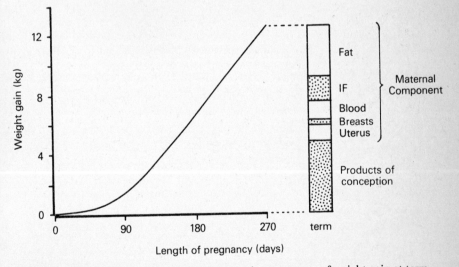

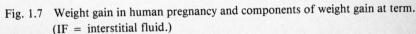

Fig. 1.7   Weight gain in human pregnancy and components of weight gain at term. (IF = interstitial fluid.)

sively in weight throughout pregnancy, showing gains of around 1 and 0.4 kg respectively by term. Uterine growth is due primarily to hypertrophy of the muscle cells, although hyperplasia may also occur early in gestation. The major stimulus to myometrial growth is uterine distension by the growing fetus. High circulating concentrations of oestrogen and proges- terone also promote uterine growth. Growth and development of breast tissue similarly occur under hormonal influence. Oestrogens are important in stimulating growth of the mammary ducts in early pregnancy. Subse- quently, prolactin, glucocorticoids and placental lactogen bring about full alveolar development.

Extracellular fluid volume is increased by almost 3 kg (litres) by term. A 50 per cent plasma volume expansion accounts for just under half of this increase (see 1.2.1): the remainder is interstitial fluid. In women with oedema, the increase in interstitial fluid alone may approach 5 kg (litres) out of a total weight gain of 14.5 kg.

Human pregnancy is characterized by the accumulation of about 3 kg of fat by the mother. Subcutaneous fat storage, as measured by skinfold thickness, occurs predominantly around the abdomen, back and thighs. Maternal fat stores are laid down primarily during the first half of preg- nancy, and provide an energy store for the third trimester when fetal growth predominates (see 1.6.3 and 1.7).

1.6.3 *Products of conception*

The products of conception account for the remaining 4.8 kg of the total weight gain in pregnancy (Figure 1.7). Of this, the fetus contributes 3.4 kg, the placenta 0.6 kg, and amniotic fluid 0.8 kg. Fetal weight, of course, may deviate considerably from this 'standard' value, and birth weight is affected by a number of variables. There is a tendency, for example, for the birth weight of first babies to decline with increasing maternal age. Fetal size also varies directly with maternal size, and some cultural differences are apparent. Indian, Oriental and African women, for example, have slightly smaller babies than their European counterparts, although this may be due in part to maternal nutritional inadequacies, particularly in the latter group. A reduction in birth weight also accompanies fetal hypoxia. This may occur in pregnancy at high altitude (see 4.6.3) and as a result of the continued use of narcotics and cigarettes during gestation.

Figure 1.8 illustrates the changing pattern of fetal growth throughout pregnancy, and contrasts it with placental and amniotic fluid changes. Thus the fetal growth curve is sigmoidal, with the greatest increase during the third trimester and a slowing down just before term. The placenta, however, shows a more gradual weight increase, without the final growth spurt seen in the fetus, although its growth is similarly retarded towards term. Changes in amniotic fluid volume parallel those described for the placenta until immediately before term, when the volume falls.

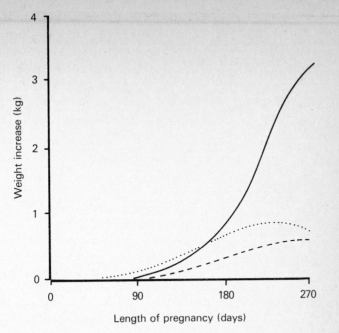

Fig. 1.8 Weight increases of the products of conception throughout pregnancy. Symbols: ——, fetus; – – –, placenta; . . . ., amniotic fluid. From Hytten, F. & Chamberlain, G. (1980). *Clinical Physiology in Obstetrics*. Blackwell: Oxford.

1.7 Metabolic changes

The altered dietary intake and gastrointestinal changes of pregnancy described earlier (see 1.5) are accompanied by characteristic alterations in several aspects of metabolism. Such changes are adaptive inasmuch as they provide for the developing fetus and also protect the mother against major nutritional deficiencies in her diet.

1.7.1 Carbohydrate, protein and fat metabolism

Glucose homeostasis is altered in human pregnancy. Fasting plasma glucose concentration falls during the first trimester by around 12 per cent from normal values of 4.5 mmol$\cdot$litre^{-1}. This lower value is then maintained until term. The reason for the early reduction in maternal plasma glucose concentration is not clear, since dietary deficiencies or excessive glycosuria (see 1.3.2.2.) are unlikely to be significant problems for most women. The increased plasma volume of early pregnancy, however (see 1.2.1), will be a contributory factor.

Towards the later stages of pregnancy, the increased glucose requirement of the fetus leads to a continuous drain of glucose from the maternal circulation. Glucose is used by the fetus primarily as a source of energy, but also to make its own fats, since some maternal fats cannot cross the placenta. The mother compensates for this loss of glucose by increasing her dietary intake (see 1.5.1.2) and also by an enhanced gluconeogenesis. During pregnancy, amino acid catabolism is reduced as glucogenic amino acids are used preferentially to make glucose. Plasma amino acid concentration consequently falls as a result of the combined effects of gluconeogenesis, fetal usage and increased renal excretion (see 1.3.2.2).

In contrast to the changes in fasting plasma glucose concentration, there occurs in pregnancy a progressive reduction of glucose tolerance following a glucose load. A standard 50 g of glucose administered orally to a pregnant woman takes a longer time to produce a maximal plasma concentration. However, the final plasma concentration is higher than normal and remains elevated longer. This may be beneficial to the fetus, since it will allow more time for placental exchange to occur. The impaired glucose tolerance in pregnancy occurs despite an increased insulin secretion, implying a decreased sensitivity of target tissues to the normal hypoglycaemic action of the hormone. This may be the result of some of the other major endocrine changes of pregnancy and both oestrogen and placental lactogen have been implicated.

Apart from altering her carbohydrate metabolism, the pregnant woman also 'spares' glucose for the fetus by switching over to using fat as her primary energy source early in gestation. Such a change precedes the requirements of the fetus, and thus represents a change in maternal metabolism in preparation for subsequent fetal demands. This alteration in metabolism is facilitated by maternal storage of fat during the first half of pregnancy. Indeed, over 3 kg of accumulated fat contributes 25 per cent of the total weight gain of pregnancy (see 1.6.2).

Fat mobilization is also enhanced, and the inhibitory effect of placental lactogen on insulin action is thought to be relevant here. One role of insulin is to facilitate the uptake of glucose by adipose tissue and subsequent deposition of fat. This is diminished during pregnancy, so that the balance of fat metabolism is altered in favour of fatty acid mobilization. Plasma free fatty acid concentration therefore increases in pregnancy, and the mother enters the final stages of gestation with a considerable energy supply which can compensate for any dietary shortcomings during the final growth spurt of the fetus.

1.7.2 *Calcium metabolism*
Fetal calcium requirements during the final trimester of pregnancy approach 0.3 g per day, just under half of the recommended dietary intake of the non-pregnant woman (Table 1.2). Pregnant women are thus encouraged to increase their calcium intake by some 70 per cent to offset the demands of the fetus (again, see Table 1.2). This fetal drain on calcium,

together with the normal haemodilution of pregnancy, cause a decrease in maternal plasma calcium concentration. Circulating PTH concentration consequently rises, and this is accompanied by an enhanced calcium absorption from the gastrointestinal tract through enhanced activation of vitamin D. Indeed, intestinal calcium uptake is doubled by the end of the second trimester.

In addition to a raised calcium absorption by the gut, urinary calcium losses are reduced following an initial increase. Stimulation of renal calcium reabsorption by PTH may account for this. The changes in calcium metabolism during pregnancy normally allow the mother to satisfy the demands of her developing fetus while preserving the integrity of her own skeleton.

1.8 Further reading

Aebi, H. & Whitehead, R. (eds) (1980). *Maternal Nutrition during Pregnancy and Lactation*. Hans Huber: Bern.

Davison, J.M. & Noble, M.C.B. (1981). Serial changes in 24 hour creatinine clearance during normal menstrual cycles and the first trimester of pregnancy. *British Journal of Obstetrics and Gynaecology* **88**, 10–17.

Findlay, A.L.R. (1983). Pregnancy. In: *Reproduction and the Fetus*, Chapter 6. Edward Arnold: London.

Hytten, F. & Chamberlain, G. (eds) (1980). *Clinical Physiology in Obstetrics*. Blackwell: Oxford.

Hytten, F. & Leitch, I. (1971). *The Physiology of Human Pregnancy*. Blackwell: Oxford.

Chapter 2

The fetus and neonate

Summary

Development of a mature human from a single fertilized egg involves many complex processes. From the time of conception, when the fertilized ovum is nurtured by secretions from the mother's oviducts, to the birth of a baby which is capable of independent existence, many organs are developed and their functions mature. Embryology deals with organogenesis and differentiation of form; this chapter is concerned with changes in function, particularly those changes that occur around birth.

When the ovum divides and forms a blastocyst, the placenta, a complex organ with no adult counterpart, begins to form. All through fetal life the placenta functions to supply the fetus with nutrients and oxygen, to rid it of waste products and to maintain ionic balances. In addition, it supplies many of the hormones on which a successful pregnancy depends.

Thus, prior to birth, the placenta functions as lungs, kidney and gut. After delivery, the neonate depends on these organs and they have to function efficiently immediately. As a consequence of the changes in function of the lungs at birth, the circulation also undergoes tremendous changes; several channels which were open in the fetus are closed. Each of these organs in dealt with in turn with respect to its function in the fetus, the changes that occur at birth, and how adult function is attained. Temperature regulation is dealt with in a similar manner.

There is no general discussion about some parts of the endocrine system in the fetus, nor of the development of the musculoskeletal or nervous systems. Development of these is progressive and birth is merely an incidental signpost, not a changing of ways; as such they fall outside the scope of this chapter.

2.1 Introduction

2.1.1 *Aims*

The fetus is a very different animal from the newborn baby. It begins life in an ideal environment with the mother being responsible for maintaining oxygen supply, nutrition, excretion and temperature regulation. Since the fetus develops in a watery medium, there are few problems of fluid balance.

The purpose of this chapter is to discuss some of the major ways in which the physiology of the fetus is different from that of an adult, and to show how the totally dependent fetus is converted into an independent organism at birth. The development of organs and embryological topics are not discussed; for these, students should consult any one of the excellent embryological textbooks available, e.g. *Medical Embryology*.

We will also consider some of the problems that the neonate has to contend with when it must fend for itself. Although examples of neonatal problems are quoted, these should not be regarded as all-embracing, rather as illustrating certain principles.

2.1.2 *Early nutrition of the fertilized ovum*

Fertilization of the ovum usually occurs in the oviduct and over the next 4–6 days the fertilized egg(s) is transferred to the body of the uterus. The ovum is maintained in a nutrient fluid in the oviduct. During the follicular and luteal phases of the menstrual cycle, oviductal fluid amounts to less than 1 ml each day; this increases to about $20 \, \text{ml} \cdot \text{day}^{-1}$ in mid-cycle. The fluid is derived from secretions of the cells lining the oviduct and from a transudate of plasma. It has a composition similar to that of plasma, although potassium concentration is much higher and calcium, protein and glucose concentrations are reduced in oviductal fluid; it also contains several unusual proteins and an increased amount of amylase and lactic dehydrogenase. Although fertilization of the ovum takes place in this oviductal fluid, *in vitro* fertilization, which has been practised for many years in a variety of animals, can occur in relatively simple solutions.

While the fertilized ovum is developing and dividing, it is nourished, as was the ovum, either in the oviduct or in the body of the uterus, by oviductal or uterine secretions. For development of the fertilized ovum, pyruvate and lactate have to be added as energy sources to simple *in vitro* solutions (since glucose can only be utilized after cleavage has commenced), but once cell division occurs, they are not necessary. For over 90 years, the later stages of development, such as blastocysts, can be and have been grown in culture, although it is only recently that human blastocysts have been cultured.

2.2 The placenta

2.2.1 *Introduction*

The outer cells of the developing blastocyst form the trophoblast, which invades the thick luteal endometrium, constituting the earliest connection with the mother and beginnings of the placenta. From this stage the uterine environment is less important as the fetus begins to derive all its nutrients from maternal sources.

The placenta is a fetal organ that has no adult counterpart. In humans its finger-like projections are bathed in maternal blood; very few animals have this type of placentation and so experimental results are often difficult to extrapolate to the human situation.

Maternal blood flow to the placenta is about 500 ml·min^{-1} towards term; it increases in parallel with the blood flow to the uterus. There is no evidence that blood flow is regulated, even though the placenta produces renin. Perhaps one reason for this lack of regulation is the destruction by the trophoblast of the muscular walls of the spiral arteries leaving merely fibrous channels.

The placenta forms a barrier between maternal and fetal blood circulations. Early in its development this barrier may be 25 μm thick and consists of the endothelial lining of fetal capillaries, mesenchymal stroma of the villus, the cytotrophoblast and the syncytiotrophoblast. As pregnancy progresses, this barrier becomes much thinner, especially the layers derived from the trophoblast, and at term the total thickness is reduced to about 2 μm. The placenta has to perform several functions for the fetus; in regard to gaseous exchange it functions instead of the lung; in regard to homeostasis and excretion it functions instead of the kidney; in regard to nutrition it functions instead of the gastrointestinal tract. In addition, both in its own right and in conjunction with other organs in the fetus, it is a major endocrine organ. It is one of the determinants of fetal growth (see 2.8.2).

2.2.2 *Placental gas exchange*

Most gases diffuse easily across the placenta, driven by a difference in their partial pressures. Carbon dioxide is typical in this respect. Hyperventilation by the mother (see 1.4.2) leads to a reduced partial pressure of carbon dioxide (PCO_2) of about 26–30 mmHg (3.5–4.0 kPa) in the maternal blood; this can be compared with a fetal PCO_2 of about 40–44 mmHg (5–5.5 kPa). These differences in PCO_2 are associated with a fetal pH about 0.12 units lower than that of the mother.

The placenta is not as permeable to oxygen as it is to carbon dioxide. A greater gradient is therefore required to transfer similar amounts of oxygen. Maternal PO_2 is about 90 mmHg (12 kPa); fetal PO_2 in the umbilical vein is maintained at 23–38 mmHg (3–5 kPa). Higher fetal PO_2 cannot be achieved because it leads to constriction of the umbilical vessels, which limits fetal blood flow to the placenta, reducing further oxygen transfer. Thus the fetus has to obtain its oxygen (at the end of pregnancy 7 mmol · min^{-1} (16 ml · min^{-1}) with 1.7 mmol · min^{-1} for the placenta) at low PO_2. Transfer of oxygen at such a low PO_2 is helped in the fetus because haemoglobin is modified. Adult haemoglobin (HbA) has two α- and two β-chains joined to form a tetramer. In the fetus two β-chains are replaced by two γ-chains which have a slightly different amino acid composition, and this is called fetal haemoglobin (HbF). HbF has a much greater affinity for oxygen than HbA, probably because it is insensitive to the the effects of 2,3-diphosphoglycerate (2,3-DPG), which in HbA inhibits binding of oxygen. So, in the fetus, the oxyhaemoglobin dissociation curve is shifted to the left. Transfer of oxygen across the placenta is also assisted by a Bohr effect (of carbon dioxide on the dissociation of oxyhaemoglobin) at both sides of the placental membrane.

Other gases, including gaseous anaesthetics, can cross the placenta. One of the most important is carbon monoxide. In acute carbon monoxide poisoning, the mother is likely to die before the fetus is severely affected, but if the fetus is subjected chronically to small doses of carbon monoxide (as happens if the mother smokes cigarettes), carboxyhaemoglobin is produced and the concentration in the fetus is twice that in the mother. As a consequence, the abililty of the fetus to carry oxygen is reduced, fetal growth is inhibited and the incidence of fetal abnormalities increase.

2.2.3 *Transfer of nutrients and electrolytes*

The placenta is the only source of supply of carbohydrates, fats and proteins for growth and metabolism in the fetus. Glucose is the major carbohydrate transported and there is a favourable gradient for diffusion to the fetus. However, transport is stereospecific in that D-glucose is preferred to L-glucose and it is thought that a carrier-mediated system is involved. In general the mechanism exhibits properties similar to those found in adult gut or kidney, but it is not certain whether placental transfer can be saturated. The placenta itself uses glucose as an energy-producing substrate and stores it as glycogen.

Free fatty acids, especially short chain ones, and glycerol diffuse easily across the placenta, and the permeability of other lipids is related to their fat solubility. Some very large lipids and lipoproteins cross only slowly. More complex lipids needed by the fetus are usually synthesized either in the placenta or in the fetus.

Proteins in general do not traverse the placenta; one exception is the IgG immunoglobulin class, which appears to be transported by a highly specific pinocytotic mechanism. Amino acids cross the placenta and achieve a concentration in the placenta and in the fetus much higher than in maternal blood. There seem to be a number of transport systems for different classes of amino acids, with L-forms being preferred; the mechanisms are similar to those in adult gut or kidney. Glutamine is used by the placenta to transfer amino nitrogen to the fetus, glutamic acid returning to the placenta where it is reconverted to glutamine. It is not known as to which amino acids are essential for human fetal development.

Water-soluble vitamins are actively transported across the placenta by highly complex mechanisms. Fat-soluble vitamins may cross by diffusion, although there is a considerable time lag between changes in maternal concentrations and the subsequent change in the fetus.

As the fetus grows, there is a need for increasing amounts of sodium, potassium, phosphate, iron, calcium and other minerals. The fetus must be in positive mineral balance. Many of these ions can diffuse rapidly across placental membranes, or even chorionic membrances, but net transfer of most ions is by specific ion pumps; many of the details are not yet known. Iron is very rapidly transported from maternal transferrin, perhaps to specific receptor sites on the placental surface; from thence it is actively transported to the fetus. Calcium, phosphate, iodine and chromium all appear to be actively transported. The placenta is not an effective barrier to some of the toxic heavy metals: mercury passes easily and lead is only slightly excluded; cadmium and ^{90}strontium, however, appear to be effectively barred from the fetus.

2.2.4 *Endocrine function*
The placenta is a source of both peptide and steroid hormones. The peptide hormones human chorionic gonadotrophin (hCG) and human placental lactogen (hPL) are secreted almost exclusively into the maternal circulation. Briefly, hCG maintains the maternal corpous luteum which secretes oestradiol and progesterone during the first 3 months of pregnancy; the function of hCG in later stages of pregnancy is not known. It is secreted by the blastocyst, possibly before implantation occurs, and rapidly reaches very high concentrations in maternal plasma (see Figure 2.1). Radioimmunoassay of this hormone is used as a diagnostic test for pregancy. Human placental lactogen is structurally similar to growth hormone and prolactin and has similar actions in that it mobilises free fatty acids, antagonises the peripheral actions of insulin and causes retention of nitrogen, potassium and other minerals. Secretion occurs from about the fourth week after conception and increase as pregnancy progresses (see Figure 2.1).

Steroid hormone synthesis requires the co-operation of placenta and fetus, since neither has the full complement of enzymes needed for their manufacture. The placental secretion of oestriol and progesterone supersedes that of the corpus luteum and increases until just before parturition. After the 12th week of pregnancy, the placenta secretes sufficient oestriol and progesterone to allow normal pregnancy to proceed in the absence of a

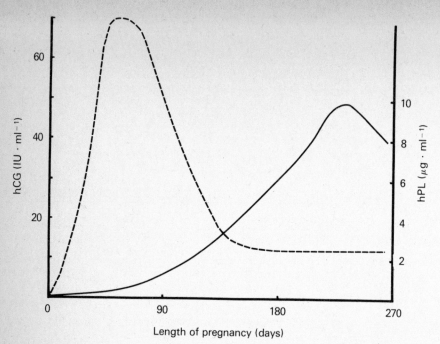

Fig. 2.1   Maternal plasma concentrations of human chorionic gonadotrophin (hCG
–––) and human placental lactogen (hPL ——) throughout pregnancy. (The length
of pregnancy is calculated from the date of conception.)

corpus luteum. For full details of synthesis and functions, readers should
consult textbooks on endocrinology.

What is surprising about the endocrine secretion of the placenta is that, as
yet, no control system has been discovered.

2.2.5 *Placental insufficiency*

If the placenta does not function normally, it can limit fetal growth, or even
cause fetal death. There may be either fetal or maternal causes for placental
dysfunction. Fetal causes include multiple pregnancies, transplacental
infections and malformations; maternal causes usually result from
restriction of blood flow or oxygen delivery to the placenta. Intervillous
blood flow may decrease because of constriction of the spiral arteries
supplying the blood (due to pre-eclamptic toxaemia, essential hypertension
or chronic nephritis) or because of heart disease. Anaemia may severely
restrict oxygen delivery.

Tests for placental insufficiency depend on assessing either growth of the
fetus (on clinical examination or using ultrasonic scanning) or endocrine
output (from the placenta). This latter method is theoretically easy, but it is

difficult in practice since the day-to-day excretion of a hormone may vary by ± 50 per cent of the mean. Analysis of 24-h urine samples over several days is a much safer indicator of function than single samples of urine or plasma.

2.3 Respiratory system

2.3.1 *Fetal respiratory movements*
Throughout most of pregnancy the fetus makes respiratory movements. Initially, these are very irregular but they become more regular as gestation proceeds. Late in pregnancy, the movements correlate well with intrauterine fetal EEG measurements. Rapid respiratory movements (60 min^{-1}) are associated with EEG patterns which in adults and children are characteristic of rapid eye movement (REM) sleep; in the intervening periods there are slower movements or perhaps only single 'gasps'. Towards term, respiratory movements occur for about 50–90 per cent of the time.

Since the fetal lung contains only a viscous secretion from the alveoli and amniotic fluid, these movements serve only to move a little fluid into and out of the upper respiratory passages; the gas exchange functions of the lung are subserved by the placenta. The amount of fluid in the lung is about the same as the functional residual capacity immediately after birth, i.e. some 30 ml · kg body weight^{-1}.

The function of the respiratory movements is not clear. The isometric work they perform may help to prepare the respiratory muscles for their work at birth; alternatively, they may be instrumental in increasing the size of the lungs during gestation. Whatever their function, the movements correlate well with the electrical activity of respiratory neurones in the medulla oblongata and indicate the development of neural control of respiratory movement.

After about 24 weeks of fetal life, type II pneumocytes in the lungs begin to secrete surfactant. The secretion rate increases towards term, particularly from 32 weeks onwards. Surfactant is really a group of phospholipids that lower the surface tension of air–liquid interfaces and decrease the surface tension as alveoli become smaller. This latter action helps to stablilize the lungs by preventing small alveoli emptying into larger ones. Surfactant serves no useful function in fetal life but secretion of adequate amounts is necessary for normal breathing after birth. Surfactant secreted into lung fluid eventually enters amniotic fluid, where its presence may be used diagnostically to determine whether a fetus is sufficiently mature to be able to maintain an independent existence.

2.3.2 *Breathing at birth*
At birth the newborn infant has to undergo an abrupt transition, with oxygen delivery being switched from the placenta to the lungs. During

passage of the fetal thorax through the birth canal, much of the liquid present in the lungs is squeezed out through the mouth or nose. The remainder is reabsorbed over the next few hours, either into pulmonary lymphatics or directly into the pulmonary circulation.

At birth the baby usually gives a gasp, takes a very deep breath and gives a cry. The normal respiratory rhythm is established within a few minutes. The first breath is usually of 30–40 ml of air and is associated with a intrathoracic pressure 40–100 cm H_2O (4–10 kPa) below atmospheric pressure. Some investigators have suggested that this negative pressure is necessary to open the lungs initially (just as an increased pressure must be used to blow up a new ballon) but this has been disputed; pressures greater than normal are needed because the compliance of the lungs (volume change for a unit change in pressure) is low.

The reason why a baby takes its first breath is not known. When the baby is delivered, it experiences many sensory stimuli for the first time. This increases neural activity in the reticular formation of the brainstem and probably stimulates breathing. Frequently the sensory stimuli are increased by the midwife either sucking mucus from the pharynx, and stimulating receptors at the back of the pharynx, or by slapping the baby's bottom. The hypoxia that develops even during a normal delivery and the concurrent mild acidosis also serve to stimulate chemoreceptors which become effective at birth.

Breathing movements in the fetus are almost entirely under the control of the nervous system. Over a period of a few minutes after birth, however, the chemoreceptors, both central and peripheral, become effective and exercise much more control over respiration. There is some evidence that the effectiveness of peripheral chemoreceptors is related to an increased blood flow, which is controlled by the sympathetic nervous system.

2.3.3 *Respiration after birth*

In the normal baby, breathing rapidly assumes a regular pattern after birth. Within a few minutes the functional residual capacity has increased to 20 ml kg · body weight^{-1} and within a few hours it has increased to as much as 30 ml · kg^{-1}. At this time the tidal volume is about 20 ml (i.e. 5–6 ml · kg^{-1}) and the respiratory rate is rapid at about 30 min^{-1} so that the total minute ventilation is 500–600 ml · min^{-1}. It is difficult to measure lung volumes in babies but advantage has been taken of periods of crying to measure vital capacity and lung capacity; the former is about 30 ml · kg^{-1} and the latter about 60 ml · kg^{-1}. Physiological deadspace is about 1.5–2 ml · kg^{-1} in infants which is 30–35 per cent of normal tidal volume — a proportion similar to that seen in adults.

Elastic recoil of the lungs is low in the newborn infant and this may result in gaseous trapping and mild hypoxia which is found after birth. The very

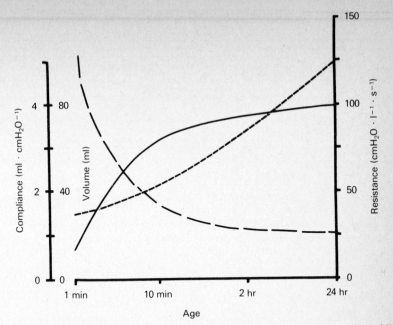

Fig. 2.2 Changes in mechanical properties of the lungs during the first day of life. Note that the horizontal axis has a logarithmic scale. Symbols: ——, resting lung volume; – – –, compliance; —— , resistance. From Godfrey, S. (1981). In: *Scientific Foundations of Paediatrics*, 2nd edn, eds J.A. Davis & J. Dobbing, Chapter 21. Heinemann: London.

low compliance of the lung in the newborn increases dramatically over the next few hours; at the same time resistance falls (see Figure 2.2).

Gas exchange across the lungs does not occur until birth and nothing is known about the way in which gas transfer develops in the newborn. Although the newborn only has about 10 per cent of the adult number of alveoli, gas exchange can also occur across the respiratory bronchioles.

Oxygen usage by the newborn infant is about 7 $ml \cdot kg^{-1} \cdot min^{-1}$, providing the baby is in a thermoneutral environment (see 2.5.2), which is about twice that in adults. Changes in environmental temperature can alter oxygen usage dramatically. Because it is difficult to make measurements in steady-state in infants, and because brown adipose tissue is metabolized at temperatures below the thermoneutral zone (see 2.5.2.1), meaningful measurements of the respiratory quotient (RQ) are difficult to make. The best estimates suggest that RQ is about 0.7 at birth and has risen to 0.8 by the end of the first week. Carbon dioxide handling is normal despite the fact that, in the newborn, carbonic anhydrase concentration is only 20 per cent of the adult value.

After birth the relative hypoxia of the fetus is corrected within 5 min,

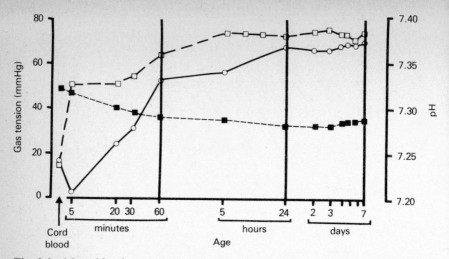

Fig. 2.3   Mean blood gases during the first week of life. Note that the horizontal axis has a logarithmic scale. Symbols: ○, pH; □, PO_2; ■, PCO_2. From Godfrey, S. (1981). In: *Scientific Foundations of Paediatrics*, 2nd edn, eds J.A. Davis & J. Dobbing, Chapter 21. Heinemann: London.

although the PO_2 continues to rise slowly over the next 5 h or so. Little change then occurs over the next 2 weeks (see Figure 2.3). The saturation of haemoglobin with oxygen rises rapidly from 40–50 per cent in the fetus to 80–90 per cent in the newborn. Changes in PCO_2 are slower and although the most rapid changes occur in the first 20 min after birth, the PCO_2 continues to fall for 2–3 days, usually falling below adult levels but regaining them gradually. The pH of blood rises more slowly than the changes in blood gases would suggest, probably because the acidosis in the fetus is metabolic and blood lactate levels are elevated. There is a rise of pH over the first week.

2.3.4 *Pulmonary blood flow*

Blood flow to the lungs depends on the PO_2 of the air in the alveoli. In the fetus the PO_2 in the fluid in the lungs is low and as a result the pulmonary vessels are constricted. Because of this increased resistance to flow, the walls of the pulmonary artery are thicker than in adults. Combined with the constriction of the vessels this results in a very high ratio of wall thickness to diameter. After the fetus takes the first breath, the increased PO_2 in the lungs reduces the resistance and increases the flow through the pulmonary vessels. Over the next 4 months there is an increasing dilatation of pulmonary arterial vessels and muscular development lags behind so that at 4 months of age the adult proportions are attained. Thereafter muscular development of the wall proceeds at the same rate as growth of the artery and the wall is a constant fraction of the total diameter.

2.3.5 *Problems with breathing*

Problems with breathing at birth and shortly after are related mainly to mechanical obstruction, asphyxia during birth or lack of surfactant. Mechanical obstruction by mucus can give problems with breathing but respiratory movements are present, at least initially, and the diagnosis and cure should be obvious.

Asphyxia can present greater problems. Even during normal delivery the fetus undergoes a mild degree of hypoxia. If this is prolonged (due to compression of the umbilical cord, prolonged labour, placental separation or any other cause), the normal pattern of breathing is changed. After an initial gasp, if there is one, the baby enters a period of primary apnoea when any stimulus may initiate gasping. If, during these gasping movements, oxygen enters the lung, the period of apnoea is terminated; if not, respiratory movements eventually cease and the baby goes into secondary or terminal apnoea. Resuscitation with positive pressure respiration is recommended and this, by forcing oxygen into the lungs, enables normal respiration to be established. In many infants the problem is made worse by sedatives which the mother has received during labour.

Because the neural mechanisms for maintaining respiration are less well developed in premature infants, there is a danger of spells of apnoea even after birth. These normally occur during REM sleep and have been suggested as a cause of unexplained cot deaths; very recent evidence does not support this view. Apnoea sometimes occurs during feeding in the early neonatal period and it is followed by a period of cyanosis. This never occurs with breast-fed babies but may occur in up to 50 per cent of those who are fed by bottle. The remedy is simply to remove the bottle and encourage the baby to belch (colloquially, 'winding').

Lack of surfactant causes respiratory distress syndrome (hyaline membrane disease). This occurs most frequently in premature babies and in babies born to diabetic or prediabetic mothers. Because the surface tension of the alveoli is unusually high, the mechanical effort required to breathe is excessive and collapse of alveoli occurs even after they have been opened by the first breath. Such large negative intrapleural pressures are generated that the intercostal muscles may be drawn inwards. The net result is that insufficient oxygen enters the lungs. Sprays containing surfactant have been used in therapy but, usually, the treatment is supportive until sufficient surfactant is secreted by the infant.

2.4 Cardiovascular system

2.4.1 *The circulation in the fetus*

The fetus uses the placenta for gas exchange and to gain its nutrients from the mother. Because of this, the fetal circulation has to be specially adapted so that blood is shunted past the lungs and special arrangements are made for the blood supply of the placenta (Figure 2.4).

Blood entering the right atrium from the superior vena cava tends to pass through the tricuspid valve into the right ventricle. Part of the blood returning via the inferior vena cava passes from the right atrium, through the patent foramen ovale to enter the left atrium; the remainder is deflected from the atrial wall, mixes with the blood from the superior vena cava and enters the right ventricle.

Blood in the right ventricle is pumped into the pulmonary artery, but only about 15 per cent finds its way eventually to the lungs. Most passes through the ductus arteriosus, a wide muscular arterial channel connecting the pulmonary artery with the aorta. The small amount of blood that passes

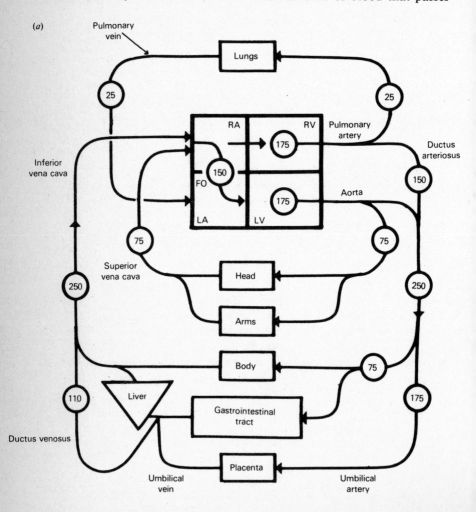

(*b*)

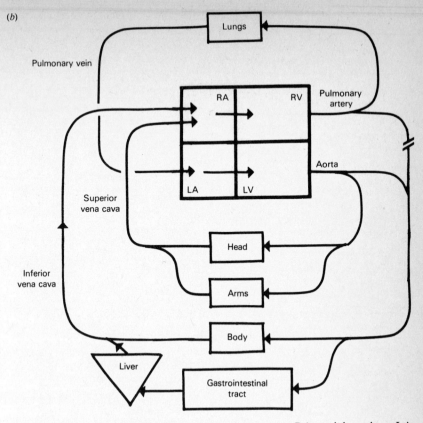

Fig. 2.4   Circulation in (*a*) the fetus and (*b*) the neonate. RA = right atrium, LA = left atrium, RV = right ventricle, LV = left ventricle, FO = foramen ovale. The figures given in (*a*) represent approximate blood flows in $ml \cdot kg^{-1} \cdot min^{-1}$.

through the lungs enters the left atrium to join with the blood delivered through the foramen ovale; this blood then passes into the left ventricle and is pumped into the aorta. Some of this blood is directed to the head and upper limbs. The remainder passes into the descending aorta and is joined by the blood entering the aorta through the ductus arteriosus; together this is distributed to the remainder of the body and the placenta.

Blood returning from the placenta flows down the umbilical vein into the left branch of the hepatic portal vein. About 40 per cent goes through the liver but the remaining 60 per cent passes through the ductus venosus, a short venous channel joining the portal vein to the inferior vena cava.

Thus, in the fetus, the two sides of the heart work effectively in parallel, not in series as in the neonate and adult (compare Figures 2.4*a* and 2.4*b*).

Because of these anatomical shunts, the PO_2 is not the same in all parts of

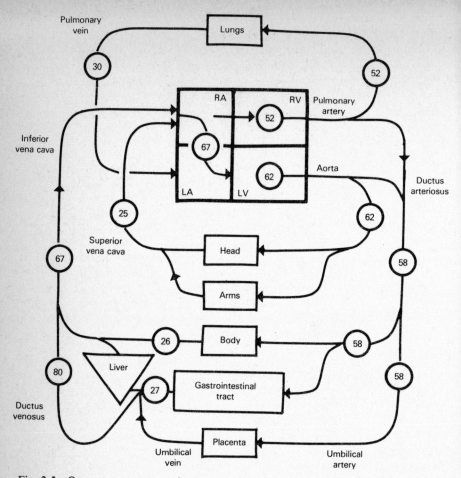

Fig. 2.5 Oxygen content at various parts of the fetal circulation. The figures given represent the saturation of haemoglobin in per cent.

the arterial system (see Figure 2.5). Oxygenation occurs in the placenta, and blood in the inferior vena cava, after it has been joined by the ductus venosus, has a high PO_2 of 26–30 mmHg (3.5–4 kPa) and is 60–70 per cent saturated. The distribution of blood flow ensures that the blood supplied to the head, upper limbs and heart contains the most oxygen. Blood supplied to the lungs has a lower PO_2 since it is derived from the right ventricle, which contains a mixture of oxygenated blood from the inferior vena cava and reduced blood from the superior vena cava. The blood supplied to the rest of the body has a lower PO_2 than that supplied to the head and upper limbs, since oxygenated blood in the descending aorta is mixed with partially oxygenated blood from the ductus arteriosus.

The right atrium receives far more blood than the left and the pressure within it is higher. This means that the pressure gradient is from right to left atrium in fetal life, which prevents closure of the foramen ovale by the septum secundum. The right and left ventricles are approximately the same thickness and generate approximately the same pressures in the aorta and pulmonary artery. However, since the peripheral pulmonary vasculature is constricted (see 2.3.4) and the placenta has a low resistance to flow, the blood flow is directed away from the lungs and to the placenta.

2.4.2 *Changes in the circulation at birth*

When the baby takes its first breath, the vascular resistance of the lungs falls dramatically. This is partly due to stretching of the alveoli, which in turn stretch the capillaries, and partly due to the effects of the increased PO_2 in the alveoli on the pulmonary vasculature (2.3.4). The net result, however, is a reduced pressure in the pulmonary circulation and, as a result, blood supply to the lungs is increased and much more blood returns to the left atrium. At the same time the reduction in pulmonary artery pressure causes the pressure gradient in the ductus arteriosus to be reserved and blood flows from the aorta to the pulmonary artery.

A few minutes after this, circulation to the placenta ceases. In nature, when the umbilical cord is torn or bitten, the vascular smooth muscle goes into spasm. This spasm may be intensified by the effects of circulating catecholamines or the stimulation of the sympathetic nerves which supply the intra-abdominal portions of the vessels. Usually, of course, these physiological mechanisms are circumvented in women as the cord is tied and then cut. The precise time at which the cord should be tied is still a matter of debate (see 2.4.5).

Because placental flow ceases, flow in the inferior vena cava is very much reduced and pressure in the right atrium falls. The increased flow of blood to the left atrium causes a rise in pressure and so the pressure gradient is now from left to right. The septum secundum acts as a flap valve to close the foramen ovale and is held in place by the pressure gradient from the left to the right atrium.

Although closure of the ductus arteriosus is not permanent at this stage, functional closure is complete 10–15 h after birth. The reasons for con-traction of the ductus arteriosus are not known with certainty but are probably related to the local increase in PO_2 that occurs. The muscle of the ductus arteriosus is exquisitely sensitive to PO_2; this sensitivity increases towards the end of gestation, so in premature infants the ductus arteriosus is likely to remain patent longer than in full-term babies. Another factor in normal closure seems to be the prostaglandins. Locally produced prosta-glandins and thromboxane, as well as circulating prostaglandins E_1 and E_2, are important in maintaining the patency of the ductus arteriosus during fetal life. It has, therefore, been suggested that prostaglandin inhibitors could be used to enhance closure in otherwise recalcitrant cases. Constriction may also be enhanced by the increased concentration of catecholamines present in the circulation at birth.

Little is known of the mechanism of closure of the ductus venosus. Flow continues for a few hours after birth. Although there is no clearly defined sphincter, the diameter of its opening into the inferior vena cava lessens when placental blood flow ceases and eventually closes completely. In spite of the lack of a defined sphincter or any known chemical stimulus to closure, the ductus venosus is less likely to remain patent than either of the other two shunts.

It must be emphasized that although the shunts are functionally closed within hours of birth, they are not permanently closed for some time. Usually the ductus arteriosus is premanently closed by 4 months. The foramen ovale is normally closed before this, but in a number of children and even adults a small communication persists. With closure of the shunts, the two sides of the heart are effectively in series and this configuration is maintained into adult life.

2.4.3 *Cardiac output and blood pressure*

Cardiac output has little meaning in fetal life because of the shunting of blood and is almost impossible to measure. Experiments in a number of animals have shown that systolic arterial pressure rises throughout gestation. In human neonates, however, the mean systolic pressure is still only 70–80 mmHg (9.5–10.5 kPa) at birth. Heart rate, which is the only fetal cardiovascular variable easily monitored in pregnancy, is usually 120–140 beats $\cdot$ min^{-1}. Mechanisms for altering stroke volume are not well developed and, as a consequence, the output of the heart is regulated solely by changes in heart rate. Fetal distress is signalled by a heart rate below 100 or above 180 beats $\cdot$ min^{-1}.

There is little information about fetal blood pressure during labour but heart rate is invariably monitored. Pressure on the fetal skull may cause an increased heart rate but bradycardia late in a uterine contraction usually signifies hypoxaemia and is a danger sign to the obstetrician.

Cardiac output in the infant is about 180 ml $\cdot$ kg^{-1} $\cdot$ min^{-1}, i.e. two- to threefold greater than that in adults. Systolic blood pressure, however, is relatively low at birth, 70–80 mmHg (9.5–10.5 kPa), but rises during the first 6 weeks. It is maintained at about 90–100 mmHg (12–13 kPa) until about the sixth year and thereafter rises slowly, achieving adult levels at about the 15th year. Heart rate at birth varies from 95 to 145 beats $\cdot$ min^{-1} but rises until it peaks in the first to third month somewhere between 125 and 190 beats $\cdot$ min^{-1}. Thereafter it falls, reaching 55–100 beats $\cdot$ min^{-1} at 15 to 16 years of age (see Figure 2.6).

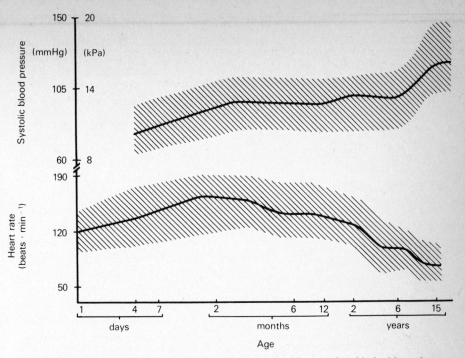

Fig. 2.6 Changes in blood pressure and heart rate with age after birth. Note that the horizontal axis has a logarithmic scale. Lines indicate mean values and the hatched area indicates values within ± 2 SD. Data from Rigby, M.L. & Shineborne, E.A. (1981). In: *Scientific Foundations of Paediatrics*, 2nd edn, eds J.A. Davis & J. Dobbing, Chapter 19. Heinemann: London.

Although the cardiac sympathetic supply is partially developed at birth, it is not as extensive as in the adult and there is considerable postnatal development. The parasympathetic supply develops somewhat earlier and reflex control of the heart at birth is dominated by parasympathetic nerves.

There is evidence from animal studies that the fetal heart can respond to sympathetic and parasympathetic transmitters halfway through pregnancy and that baroreceptor reflexes are present a little later. Chemoreceptor reflexes develop at about the same time. It has been suggested that the increased heart rate during the first year of life is due to the increasing influence of the sympathetic system. It may be that pronounced fetal bradycardia in response to hypoxaemia is dependent on parasympathetic reflexes and so may be ameliorated by selective parasympathetic blockade.

2.4.4 *Blood cells*
Production of blood cells occurs in different organs during different stage of development. Earliest development takes place within the developing

blood vessels of the yolk sac but by the sixth to seventh week of fetal life, extravascular manufacture also occurs in the liver. From the third to sixth month of fetal life, most blood cells are produced in the liver, with the spleen contributing a small amount; lymphocytes are produced in the lymph nodes and thymus gland. About the fourth to fifth month, the bone marrow begins production of blood cells. The liver is concerned mainly with red cell production and initially the bone marrow produces mainly granulocytes. During the last 3 months of fetal life, more and more red cells are produced in the marrow and at birth the marrow and lymph nodes produce most of the cells in blood.

At about the 10th week of gestation, fetal red blood cells are very much larger than their adult counterparts, with a mean corpuscular volume (MCV) of 200–250 fl. and the red cell count is in the range $1–1.5 \times 10^{12}$ litre^{-1}. As can be seen from Figure 2.7a, there is a progressive rise in haemoglobin concentration until about the 25th week of gestation; there is a further rise just before birth when haemoglobin concentration is greater than in the adult. After the 10th week, red cell count increases and MCV rapidly decreases to almost adult values. The number of reticulocytes is high (3–5 per cent of total red cells) and there are frequently nucleated red cells in peripheral blood.

The haemoglobin concentration of the blood and the haematocrit rise during the first few hours after birth. The reason for their rapid rise is not

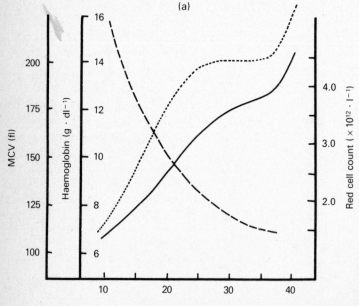

(a)

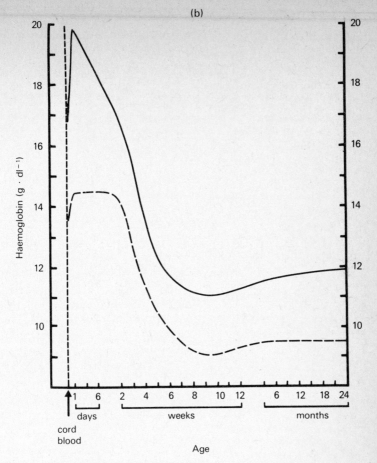

Fig. 2.7 (*a*) Haemoglobin concentration (Hb, ----), red cell count (RCC, ———), and mean red cell volume (MCV, – –) in human fetus; (*b*) haemoglobin concentration in the first two years of life, showing mean value (———) and lower limit of normal (– –). Data from Black, P.J. & Barkham, P. (1981). In: *Scientific Foundations of Paediatrics*, 2nd edn. eds J.A. Davis & J. Dobbing, Chapter 25. Heinemann: London.

certain, but is unlikely to be production of new red cells. The total number of red cells and the total amount of haemoglobin in the newborn depend initially on the amount of 'placental transfusion'. At birth up to 35 per cent of the baby's blood volume may be contained in the placenta. If, at birth, the placenta is held higher than the baby, much of the placental blood can be delivered to the bady; it has not been determined how much is optimum. Much of this 'transfused' fluid is absorbed into cells or passed out via the kidneys; the net result is an increase in haemoglobin and red cell count.

After a few days the haemoglobin concentration gradually falls (Figure

2.7b). This physiological anaemia is present regardless of the amount of 'placental transfusion'. It is explained in the following way. Because of low PO_2 in the fetus, the kidneys produce large amounts of renal erythropoietic factor which generates erythropoietin which stimulates erythropoiesis. This may account for the high haemoglobin concentration prior to birth. At birth, PO_2 rapidly rises and erythropoietin is not detectable in the circulation. Erythropoiesis is depressed and few new cells are formed. Erythropoietin reappears in the plasma after about the fifth week of extrauterine life and more new cells are produced, giving rise to an increased reticulocyte count (up to 2 per cent) by the ninth to tenth week; this then reduces to about 1 per cent after 3–4 months.

Less is known about white cell production. It begins about the same time as erythropoiesis but occurs mainly in the bone marrow and lymph nodes. By the 25th week the granulocytes total about 1×10^9 litre^{-1} and lymphocytes 4×10^9 litre^{-1}. By term, the granulocytes have increased to 8×10^9 granulocytes litre^{-1} whereas lymphocytes remain approximately the same; there is, however, a very wide range of all these values. The granulocyte count increases for the first 24 h after birth but falls abruptly between days 4 and 7, when it is similar to adult values. The number of lymphocytes increases after birth and reaches a maximum at about 6 months of age when there are about 7.5×10^9 litre^{-1}. After the first year of life the numbers gradually decline to reach adult values at about 16 years. Between the ages of 2 weeks and 5 years the lymphocyte is the predominant white cell.

2.4.5 Circulatory problems

Apart from congenital malformations, which are outside the scope of this book, relatively few types of problem occur in the cardiovascular system.

Overtransfusion of placental blood by draining all the blood from the placenta leads to an elevated blood volume, haematocrit and arterial pressure which may have consequences for renal function. Effectively the circulation has been overloaded. Undertransfusion can be as much of a problem, resulting in a pale baby with a low blood pressure and haematocrit. These problems are much more severe in premature infants because relatively more of the total blood volume is contained in the placenta.

Most of the problems with the circulation arise because of asphyxia, which can reverse the changes in the circulation that occur at birth. Because the pulmonary vessels are so sensitive to hypoxia and respond by partial constriction, the pulmonary artery pressure is raised by neonatal hypoxia. This reverses the flow in the ductus arteriosus so that once again blood passes from the pulmonary artery to the aorta; the pressure in the left atrium declines so that blood again flows through the foramen ovale. Thus the circulation at birth is 'brittle' and asphyxia can cause the neonatal circulation to revert to the fetal pattern. Because the lungs are the only means of oxygenating blood, this reversal of flows results in further hypoxia — a potentially fatal positive feedback.

2.5 Temperature regulation

2.5.1 *Thermoneutral zone after birth*
The fetus has no problems regulating its temperature since it is surrounded by amniotic fluid at body temperature. Maternal regulatory mechanisms are responsible for supplying or dissipating heat. At birth, though, dramatic changes in environmental temperature occur as the child is thrust into a cold, cruel world. Although the modern obstetrical delivery unit is maintained at a high temperature, it is still some 8–10°C below the environment the baby has just left.

The baby is homeothermic; it defends its body temperature using a series of mechanisms for increasing heat loss and other mechanisms for producing extra heat. As described in detail for the adult in Chapter 5, heat is lost through the skin by conduction, convection and radiation and also by evaporation of sweat, all of which are critically dependent on blood flow to the skin. Heat is produced by cellular metabolism. These processes must balance to achieve homeothermy.

The thermoneutral zone (see 5.1.1) is defined as the environmental temperature range over which metabolic rate is kept at a minimum, and within which temperature regulation is achieved by non-evaporative physical processes alone. This zone is much higher in the naked neonate (32–36°C) than in naked adults (27–31°C). Clothing provides a microclimate around the body which helps to achieve this temperature.

In the thermoneutral zone, temperature is maintained by vascular changes only, but because the autonomic nervous system is not yet fully developed, the baby is not as capable as the adult of dealing with small temperature change.

2.5.2 *Special problems of the neonate*
The major problem the neonate has in maintaining body temperature is retaining body heat. As in all small mammals, the body surface area to volume ratio is high. Since loss of heat is a function of surface area, and generation of heat (i.e. metabolic rate) is a function of body mass or volume, anything that increases the ratio causes problems with heat retention. There are additional problems. Immediately after birth the baby is wet, and evaporation of the amniotic fluid can cause tremendous loss of heat. Neonates have relatively little subcutaneous fat to insulate their skin. Clothing obviously helps to protect against heat loss. What is not always appreciated is the amount of heat that can be lost from the baby's large head, whose skin is abundantly supplied with blood.

All the above problems are magnified in preterm or small infants. The smaller the baby, the greater the problems.

2.5.2.1 *Heat production*. The baby has only limited resources to combat loss of heat. Metabolic activity in all cells, but particularly in the liver and the brain, produces a fixed amount of heat. As the baby grows older his basal metabolism increases; as judged from oxygen usage it doubles in the first 10 days of life.

Voluntary muscular activity, which can generate so much heat in adults (see 5.2.1.1), accounts for little in the newborn because the neural control mechanisms for muscle are poorly developed. Likewise, shivering is rarely seen in neonates. This may be, because neuromuscular development is not sufficiently advanced to permit it, or because other mechanisms become effective before the temperature at which shivering occurs is reached. Both adults and neonates exhibit non-shivering thermogenesis; one mechanism which is very important in the neonate but is of questionable importance in the adult (see 5.2.1.2) involves brown adipose tissue.

In human neonates, brown adipose tissue is disposed as a thin sheet between the scapulae, around the neck, behind the sternum and around the kidneys and adrenals (Figure 2.8). The amount and location differ very much in different species. In the human neonate, brown fat is 2–6 per cent of the total body weight. When these stores are depleted, the baby is very much at the mercy of environmental temperatures. Although brown fat is present from the fifth month of gestation, most is laid down in the last weeks before birth and so premature infants, who probably need it most, have reduced amounts of brown adipose tissue. It is not clear how long brown adipose tissue persists into adult life but its major role in neonates is during the first 6 months of life. Brown adipose tissue has a good blood supply, a well developed nerve supply and many mitochondria placed adjacent to numerous small vacuoles filled with fat. When the vacuoles are full of fat, the tissue is yellow, but when the vacuoles are empty, the tissue is brown (hence its name) because of the high content of mitochondrial cytochrome enzymes. This contrasts with ordinary with adipose tissue whose cells have large fat droplets and the cytoplasm is compressed to a narrow rim; its blood supply is meagre.

The response of the neonate to cold temperatures is integrated in the hypothalamus and mediated through the sympathetic nerve supply to brown fat. Release of noradrenaline from the nerve terminals (or circulating adrenaline) stimulates thermogenesis by hydrolysis of stored triglycerides. Whereas, in white adipose tissue, lipolysis occurs and glycerol and fatty acids are transported to other sites for further metabolism, brown adipose tissue, because of its high content of mitochondria, is able to utilize the fatty acids and eventually convert them into energy via the tricarboxylic

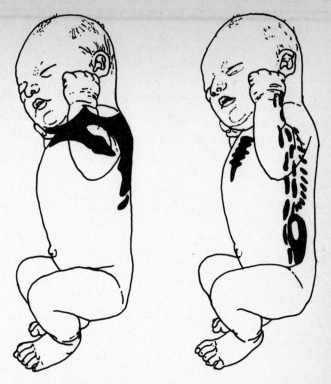

Fig. 2.8 Location of brown adipose tissue (shaded areas) in the human neonate. Major deposits are found between the scapulae and around the neck, spine and sternum. From Dawkins, M.J.R. & Hull, D. (1965), The production of fat. *Scientific American* **213**, 63. Copyright © 1965, by Scientific American, Inc. All rights reserved.

acid cycle. This produces much heat locally, which helps to maintain body temperature.

2.5.2.2 *Heat loss*. As stated previously, the major problem of temperature regulation in neonates is retention of heat. This is fortunate, since mechanisms for increasing heat loss are not well developed. If a baby has to lose excess heat because of very high environmental temperatures (in Great Britain these probably only occur artificially as in faulty incubators and in cars with all the windows closed exposed to direct sunlight on a hot summer's day), vasodilatation is of little help. This is because, when environmental temperatures exceed body temperature, heat is added to the body and this is increased by vasodilatation. Only evaporation of sweat can cause effective cooling. However, sweating is not well developed in infants. Sweat glands are present in the skin but the rise in body temperature necessary to initiate sweating (as much as 1°C) is much greater than in the adult. The number of active sweat glands reaches a maximum at about 2 years of age but many glands do not have the capacity to secrete even in adult life, although they appear anatomically normal.

2.5.3 Consequences

In spite of all the potential problems, most babies have little difficulty in regulating their body temperature. By providing adequate clothing (especially for the head) and a reasonably warm environment, parents ensure that undue amounts of brown adipose tissue are not metabolised under 'normal' conditions.

For some babies, however, there is a need to monitor their temperature more closely. Premature babies have increased problems with temperature regulation and may not be able to manage unaided. Ill babies also have increased problems because the intake of nutrients is reduced and because, for a variety of reasons, they need to be nursed naked. Under these conditions it is far easier to maintain temperature in a closely regulated environment and babies are nursed in incubators (see 8.3).

2.6 Renal function and fluid balance

2.6.1 Renal function in the fetus

The fetus contains relatively more water than the neonate or the adult. In the smallest embryos, water accounts for 92 per cent of the body weight. This decreases towards term when it reaches about 80 per cent, which is still high when compared with the 72 per cent of lean body mass which is 'normal' in children and adults. Much of this extra water is extracellular. Throughout fetal life the volume of extracellular fluid exceeds that of intracellular fluid.

During fetal life the placenta is the major organ of homeostasis and during this time the kidneys are maturing and growing. Juxtamedullary nephrons and glomeruli are the first to function and not all the cortical glomeruli are functional at birth. Even allowing for the smaller size of the fetus, renal blood flow is reduced when compared with that in the child. This probably arises partly because of incomplete development of some glomeruli and partly because of arteriolar vasoconstriction. Concomitant with this reduced blood flow, there is a reduced glomerular filtration rate (GFR).

Copious amounts of urine are produced by the fetus in spite of the low GFR. This arises because the fractional excretion of sodium (i.e. the fraction of the sodium filtered at the glomerulus that is not reabsorbed) is very high (13 per cent at 4 months of gestation, falling to about 3. 5 per cent at term) when compared with children and adults (about 0.3 per cent).

Attempts to monitor urine production in the fetus have been made using ultrasound and it is possible to 'see' the bladder fill and empty regularly; computations of urine flow rate have been made from such scans. The urine contributes a large fraction of the amniotic fluid; if the kidneys are congenitally absent, there is a great reduction in the volume of amniotic fluid (oligohydramnios). Why the kidney excretes so much sodium is not know; it may be because the kidney is immature or because there is a specific mechanism inhibiting sodium reabsorption.

Early in gestation the amniotic fluid is isotonic but as pregnancy proceeds it becomes hypotonic. This is because the maturing kidneys produce a hypotonic urine, but why they do so is unknown. Studies in animals have shown that, at least towards the end of pregnancy, the renal medulla is capable of concentrating urine, but under normal circumstances does not. Because there is an osmotic gradient between the amniotic fluid and maternal tissues, there is considerable loss of fluid across the chorioamniotic membranes. If this is not replenished by the kidneys, oligohydramnios results.

2.6.2 *Renal changes at birth*

Before birth the kidneys can excrete extravagant amounts of salt and water as deficiencies in the fetus are rapidly corrected by placental transfer. Once the umbilical cord is cut, however, fluid losses must balance fluid intake. The changes in renal function are not as spectacularly rapid as those occurring in the cardiorespiratory systems, but they are just as important. Changes occur over the first few days.

For the first few hours after birth, GFR and fractional sodium excretion are similar to prenatal values. Over the next 24–48 h both fall rapidly and then, in the succeeding 6–7 days, gradually recover as the baby takes in fluid (Figure 2.9). The extent of the natriuresis which occurs during the first few hours is modified by the size of the 'placental transfusion' (see 2.4.4). For the first few hours the baby goes into negative water and sodium balance. The major part of this loss is borne by the extracellular fluid compartment whose volume is reduced during this time until it is less than that of the intracellular compartment; this ratio is maintained throughout the rest of life. After about 7–10 days, urine production and GFR have increased and input and output of salt and water are balanced.

Plasma urea concentration decreases consistently over the first week of life as does plasma creatinine concentration. The fall of the latter can be explained by the increase in GFR which occurs after the first few hours. The decrease in plasma urea is more difficult to explain; it may just be that the kidney is more efficient than the placenta at excreting urea or that some of the urea in the fetus was derived from the mother across the placenta. Plasma osmolality is maintained constant over the first week.

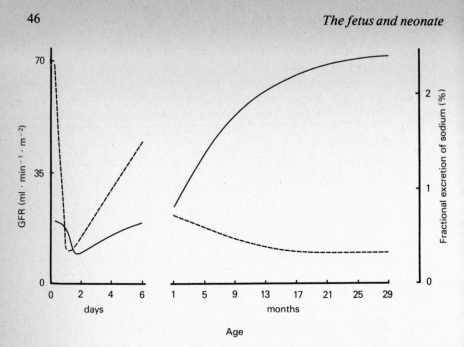

Fig. 2.9 Changes in glomerular filtration rate (GFR, ———) and fractional excretion of sodium (–––) in the neonate and child. Adult values are attained after approximately 20 months.

2.6.3 *Renal functon in the neonate*

Although experimental data are incomplete, it is clear that the neonatal kidney is still immature in many respects. Both GFR and tubular function are proportionately less than in the adult and remain so until about the second year of life.

Glomerular filtration rate rises almost exponentially from birth to achieve adult rates (when related to body surface area) after 2–3 years of life (Figure 2.9). This increase occurs for a number of reasons. First, the outer glomeruli, which are non-functional at birth, mature; secondly, cardiac output increases in absolute terms, and increases renal blood flow; thirdly, the proportion of the cardiac output which goes to the kidney increases rapidly, because renal vascular resistance is reduced much more than vascular resistance in other areas (except the lung). Both these latter changes increase renal blood flow which, in turn, increases GFR since renal plasma flow is a major determinant of GFR. Because of the decreased renal vascular resistance (which is mainly due to the afferent arterioles), the pressure in the glomerular capillaries rises; direct measurements in guinea pigs have shown that the effective ultrafiltration pressure increases by a factor of two or three in early postnatal life. Because of the maturation of the outer cortical glomeruli, progressively more blood goes to the outer parts of the renal cortex than in the fetus.

Tubular function is also less efficient in the neonate than in the adult. This is not so apparent under normal conditions, but manifests itself as a lack of reserve in dealing with excessive loads. For example, if a baby has to excrete an excessive sodium load, this takes much longer than would be expected for an equivalent load in the adult, and the baby may become oedematous. Maximum tubular reabsorption of glucose is also said to be reduced.

Tubular secretory mechanisms are depressed. Para-amino hippuric acid (PAH) is a substance that is secreted mainly by the later parts of the pars convoluta and pars recta of the proximal tubule. In adults, 85–100 per cent of plasma PAH is extracted in one passage through the kidneys, provided the plasma concentration is sufficiently low; in infants this may be reduced to as little as 50 per cent at the same plasma concentration. Since the clearance of PAH is the classical way of measuring renal plasma flow, it is difficult to measure renal plasma flow in neonates.

Decreased tubular function in neonates may simply be a consequence of the reduced lengths of both proximal and distal tubules. Whether this explains the facts fully, or whether these 'deficiencies' are really the normal physiological response to alterations elsewhere in the infant, is not known.

Babies are unable to concentrate their urine to the same extent as adults. It was thought that this was also due to functional immaturity of the nephron, but another more reasonable explanation has now been offered. Most of the protein in the baby's diet is converted to amino acids which are used to synthesize protein for incorporation into the tissues; very little is metabolized to produce energy and, as a consequence, the plasma urea concentration is relatively low. Urea is essential for the efficient functioning of the concentrating mechanism. If infants are given a high protein diet (or even urea) in their feeds, they have almost the same ability to concentrate their urine as adults.

2.6.4 *Fluid balance in the neonate*
Fluid balance in the neonate is much more precarious than in adults. This arises partly because the baby has little control over its fluid intake; most fluid is given as milk and it is difficult for the baby to communicate thirst. In addition, fluid losses are necessarily high because of increased insensible water loss from the skin (the surface area to volume ratio is high) and from the respiratory tract (the respiratory rate is high). Increased motility of the bowel, because it results in increased passage of semifluid faeces, also accounts for extra water loss.

The small size of babies represents a further problem. Small losses of fluid assume a much greater significance when subtracted from a small infant with a high water content. This can give rise to problems in babies nursed in special environments (see 8.3).

Because the kidneys cannot immediately conserve fluid at birth and because there is no fluid intake for a number of hours, there is a reduction in total body fluid volume, in extracellular fluid volume and in weight. Once fluid intake is established on a regular basis, the kidneys are able to regulate body fluid composition. However, as explained earlier the concentrating mechanism is not well developed and so there is always a danger of dehydration.

2.6.5 *Problems with fluid balance*
Because fluid balance is so precarious, it is easy to tip an infant into imbalance. Usually the problem is dehydration, although overhydration is also possible. Conditions that cause extra losses of fluid, such as diarrhoea and vomiting, predispose to dehydration, and more so since any replenishment of body fluids by mouth is precluded. Any baby that does not take its feed is liable to become dehydrated, so that problems of fluid balance frequently complicate other conditions. If parenteral routes are not appropriate, intravenous fluids need to be given.

Another problem that has been recognized in recent years is incorrect formulation of artificial feeds for babies. Although there is no danger if feeds are made up according to the instructions, a surprising number of mothers do this incorrectly. Using concentrated feeds, as well as increasing protein, fat and carbohydrate intake, increases the intake of ions such as sodium. Since these ions cannot be rapidly excreted, the plasma concentration of sodium and plasma osmolality may rise, causing retention of fluid and a rise in extracellular fluid volume. Overdilution of feeds or giving too much 5 per cent glucose in water could have the oppostie effect.

2.7 Gastrointestinal tract and metabolism

2.7.1 *Gastrointestinal function in the fetus*
The fetus obtains all its nutritional requirements from its mother via the placenta. Glucose, which supplies most of the energy, amino acids, which are used largely for protein synthesis, and fatty acids cross the placenta easily but the passage of some lipids is restricted (see 2.2.3).

Plasma glucose concentration in the fetus is about half that in the mother. Although there is some control of fetal plasma glucose concentrations (disastrously low levels are prevented), regulation is not very efficient and changes in maternal glucose concentration tend to be reflected in the fetus. Towards the end of gestation, relatively large amounts of glycogen are stored in fetal muscle and liver to succour the baby after birth, before feeding begins.

Although swallowing is established early in fetal life, other movements of the gastrointestinal tract are not well developed. Meconium (see 2.7.3.2), which is present in the gut before term, is not normally found in amniotic fluid; its presence there denotes fetal distress.

Salivary secretion, or at least salivary amylase secretion, occurs in the latter half of pregnancy. There is evidence that pancreatic enzymes are formed at about the same time. At the fourth to fifth month of gestation, gastric glands begin to appear in the stomach, but the gastric contents at birth are usually neutral.

2.7.2 *Changes in gastrointestinal function at birth*

At birth the intravenous nutrition of the fetus is suddenly interrupted, but parenteral nutrition does not begin for a number of hours. Over this period the neonate must depend on the metabolic stores laid down late in gestation. Babies born prematurely obviously have a problem because the stores are less.

Blood glucose concentration during parturition is usually higher than in the fetus. This is attributed to the high concentrations of circulating catecholamines and the increased activity of the sympathetic nervous system causing glycogenolysis.

2.7.3 *Changes in gastrointestinal function in the neonate*

There is still debate about the optimum time to give the neonate its first feed. Custom, hospital routine, convenience and various social factors seem to play as large a part as physiology in this decision. Instituting feeding has a profound effect on the gastrointestinal tract: relatively large volumes of milk need to be mixed, propelled, digested and absorbed by the gut efficiently and almost immediately. Babies born two months prematurely rapidly adapt to oral feeding, suggesting that there is an environmental trigger to the changes that occur.

At the same time the major source of energy for the neonate changes from glucose to fats and the infant has to achieve metabolic homeostasis unaided. Many of these changes are mediated by gastrointestinal hormones.

2.7.3.1 *Gastrointestinal hormones.*

There is good evidence that motilin, GIP (gastric inhibitory polypeptide or glucose-dependent insulin-releasing peptide), gastrin, enteroglucagon and pancreatic polypeptide all increase immediately after the first feed and achieve plasma concentrations much greater than those in the adult (Figure 2.10). If, for any reason, oral feeding is not instituted, these changes are much delayed. Peak basal plasma concentrations of the peptides are reached in the first or second week of extrauterine life. The increase in plasma concentration that occurs after a

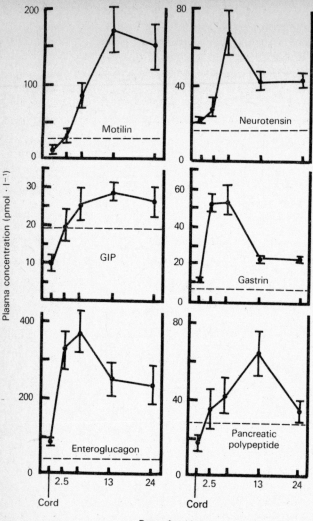

Fig. 2.10 Postnatal surges in basal plasma concentrations of gastrointestinal hormones. Each neonate gave one datum point only. Broken lines show mean adult fasting values. From Lucas, A. (1981). In: *Scientific Foundations of Paediatrics*, 2nd edn, eds J.A. Davis & J. Dobbing, Chapter 6. Heinemann: London.

meal in adults (the dynamic response) takes longer to develop and it may be 2–4 months before the baby has a 'normal' response to each feed.

The reasons for the persistently raised basal concentration are not clear. Relative to the rest of the gut, the endocrine cell mass is much greater in the neonate than in the adult and it may be that gut peptides have a special role to play in postnatal development of the gastrointestinal tract. Another contributory factor to the raised concentration is a deficiency of the mechanisms that clear the plasma of the peptides.

Most of the knowledge of the physiological actions of these peptides has been derived from studies in mature animals or humans and definitive experiments in neonates have not been performed. Nevertheless, an overall picture is beginning to emerge. Gastrin and enteroglucagon, in addition to any other actions, stimulate growth of the gastrointestinal mucosa; in the first few weeks of life this growth is rapid. Gastrin, cholecystokinin–pancreozymin and pancreatic polypeptide may stimulate growth of the exocrine pancreas. Motilin increases gut motility in adults and it is possible that it accounts for the increased motility of the whole of the gut seen after birth. Enteroglucagon and GIP are thought to be major factors in controlling insulin release.

There is obviously a complex relationship to be untangled before we fully understand these mechanisms. It is likely that different stimuli are responsible for different hormonal secretions since there are differences both in basal and dynamic secretion between breast- and bottle-fed babies.

2.7.3.2 *Gastrointestinal motility*. This has not been studied extensively. The first stool is usually passed on the first day of life and consists of a sticky, black or dark-green mass; this is meconium. It contains epithelial cells, intestinal secretions, bile pigments (which give the colour) and large amounts of proteoglycans, which are the undigested residues of the mucous secretions from all parts of the gut. Usually meconium disappears within 2–3 days and is replaced by a green–brown greasy stool, which contains much fat. This then gradually changes to the yellow, semiliquid stool of the breast-fed baby. Cow's milk, introduced into the diet in even small amounts, makes the stools much firmer. Bowel motions occur much more frequently in the neonate than in the adult, as many a harassed parent has found. There may be up to 12 stools a day, although there is great variability, and even a normal baby may occasionally go 4–5 days without passing faeces. Defaecation is reflex and full control is not usually gained until 2–3 years of life.

2.7.3.3 *Digestion and absorption*. These are basically the same as in adults, with two notable exceptions. Lactose is the major carbohydrate in the neonatal diet and this is hydrolysed by a specific disaccharidase, lactase, at the brush border of the small intestine. The amount of various disaccharidases present seems to be determined by the diet (i.e. they are inductable). The second exception is in absorption of specific proteins. Breast-fed babies have significantly raised plasma concentrations of immunoglobulins (IgG,

IgA and IgM) when compared with bottle-fed infants. These probably cross the intestinal wall by pinocytosis. When this process ceases varies from species to species, and in humans it is not known how long it persists in neonatal life. It probably plays a significant part in passive transfer of immunity from mother to child.

2.7.3.4 *Liver function.* Many of the liver's functions are similar to those in the adult though they are not as well developed. This shows particularly in two respects. The activity of the enzymes responsible for conjugation of bilirubin is low at birth but increases rapidly. This immaturity, combined with the increased breakdown of the blood derived from the placenta (see 2.4.4), results in a physiological jaundice in the first few days of life. A related aspect is that detoxification of many drugs and hormones, which usually occurs in the liver, is not well developed. Administration of drugs to the newborn can present many clinical problems.

2.7.3.5 *Energy metabolism.* Immediately after birth the baby has to depend on its own reserves of stored energy until it has access to new supplies in the form of milk. Glycogen is used for as long as supplies last, but if for any reason a neonate is not fed for 1–4 days, profound hypoglycaemia occurs. Plasma glucose concentration is normally raised at the time of birth (see 2.7.2) but then falls to about 3 mmol $\cdot$ litre^{-1} (54 mg $\cdot$ dl^{-1}) for the next 3–4 days, gradually climbing thereafter until it reaches adult levels somewhere between 1 and 6 months of age.

Free fatty acid concentrations in plasma rise rapidly after birth, the rise usually beginning in the first 2 h. Concentrations then decline over the next 12 months to reach adult concentrations. As with glucose, the increase is probably caused by an increased sympathetic activity, although immaturity of the liver may contribute. It is important to realize that cold exposure leads to hydrolysis of brown adipose tissue (see 2.5.2) whereas starvation results in hydrolysis of white fat.

2.7.4 *Gastrointestinal problems*
There are several types of problem that affect the gastrointestinal tract of the neonate. One type of problem is caused by lack of digestive enzymes. This can be a specific defect (usually congenital) in which one enzyme is missing, or a more generalized disease such as cystic fibrosis, which damages the pancreas and reduces the production or secretion of all pancreatic enzymes. The most important single gastrointestinal defect is probably lactase deficiency.

Vomiting and diarrhoea cause problems to the neonate mainly because of loss of water and electrolytes. Diarrhoea is associated with disturbances of transport across the cells of the intestine. This may be due to a defect of digestion (e.g. lactase deficiency), or a defect of absorption such as may occur in the rare congenital chloride diarrhoea, where mechanisms for chloride absorption are disturbed. If incomplete digestion is the cause, bacterial fermentation of sugars may increase the acidity and osmotic activity further and result in irritation of the bowel, with a watery, acid diarrhoea. Vomiting can be spectacular if due to pyloric stenosis or other mechanical obstruction of the tract. However, many normal babies regurgitate some food after feeding, especially if they are not 'winded' properly, largely because the lower oesophageal sphincter is not fully competent.

Hypoglycaemia can occur in infants from a variety of causes. The signs, which range from enhanced activity (twitching, a high pitched cry and occasionally convulsions) to reduced activity (limpness, difficulty in feeding and, eventually, coma), are very non-specific. What is surprising is the remarkable tolerance of the neonatal nervous system to hypoglycaemia. It seems to be able to use metabolic substrates other than glucose more efficiently than its adult counterpart.

2.8 Endocrine system

2.8.1 *Introduction*
Not all endocrine organs are considered here. Many have similar functions in both fetus and adult, but for other endocrine organs information is fragmentary. In general, peptide hormones do not cross the placenta so those occurring in fetal plasma are derived from the fetus. Some of the smaller peptides such as thyrotrophin-releasing hormone (TRH) are able to cross, however, and exert an influence on the fetus. Maternal steroid hormones cross to a variable extent, but such high concentrations of oestrogens and progesterone are present that they may have a significant effect on the fetus (see 2.2.4).

2.8.2 *Growth*
Growth of the fetus *in utero* can be assessed by ultrasonic techniques which give a more accurate overall picture than measurements derived from infants born at various gestational ages. However, since ultrasound is a relatively new technique, not all relevant measurements have yet been made. Of the various parameters (weight, length, head circumference and subcutaneous skinfold thickness) used to assess fetal growth, weight is used most consistently, but this cannot be measured using ultrasound.

The fetus does not control its own rate of growth. Although fetal plasma growth hormone concentrations are normally higher than in neonates and adults, even anencephalic fetuses, who have no hypophysis, grow at normal rates. The superabundance of human placental lactogen which is responsible for much of the maternal 'growth' is not found it fetal plasma. Fetal growth is determined by genetic factors, which limit the maximum growth that can occur, by placental function, and by the supply of nutrients from the mother. Growth is stunted if there is placental insufficiency.

After birth, the control of growth passes to the neonate, although genetic and nutritional influences are important. Weight gain accelerates over the first weeks postnatal life, reaches a maximum after 1–2 months and then slows (see Figure 2.11). Preterm babies tend to increase their postnatal growth spurt, and to maintain it for rather longer until they have 'caught up' with genetically determined growth.

2.8.3 *Adrenal cortex*

The fetal adrenal cortex has a distinct region known as the fetal zone which produces dehydroepiandrosterone (DHA) from pregnanolone produced in the placenta. This it does by splitting off the C21 side chain from the steroid

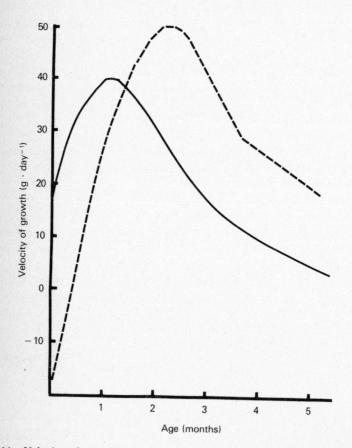

Fig. 2.11   Velocity of growth in term (——) and preterm (– – –) infants.

nucleus. This cleavage is an important step in the production of oestrogens by the placenta but may have little direct significance to the fetus. The adult zone of the adrenal cortex, which increases in size from mid-term onwards, produces glucocorticoids and aldosterone. Adrenocorticotrophic hormone (ACTH) stimulates both fetal and adult zones of the cortex but may be depressed in the fetus by maternal steroid hormones crossing the placenta and inhibiting its secretion. Melanocyte-stimulating hormone (MSH) and corticotrophin-like intermediate lobe peptide (CLIP) stimulate the fetal zone. The functions of the fetal zone cease at birth, but, for a few weeks thereafter, several steroids are present in neonatal urine which are breakdown products of fetal zone intermediates.

The adult zone is capable of producing enough glucocorticoids and aldosterone at birth to maintain life. Since the placenta regulates salt and water balance, it is unlikely that aldosterone has any significant function in the fetus, but in the neonate it acts as it does in the adult. Cortisol stimulates surfactant production late in pregnancy but whether this can be turned to therapeutic use in prevention of respiratory distress syndrome (hyaline membrane disease, see 2.3.5) is not known.

The fetal adrenal cortex is concerned with the initiation of labour in the mother. Maturation of higher centres in the fetal brain results in an increased secretion of ACTH from the fetal hypophysis. This overrides maternal suppression, and results in an increased secretion of cortisol and DHA by the fetal adrenal cortex. The increased secretion of DHA results in an increased oestriol production and a decreased progesterone production by the placenta, and one, both, or all three of these changes eventually result in an increase in the concentrations in uterine prostaglandins, which initiate labour.

2.8.4 *Adrenal medulla*

The fetal adrenal medulla initially produces exclusively noradrenaline but the proportion of adrenaline increases as gestation proceeds. The function of the catecholamines secreted by the fetus is not clear although it has been suggested that they constrict the umbilical and placental blood vessels. During delivery their major function is probably to prevent hypoglycaemia by stimulating glycogenolysis. In the neonate, plasma noradrenaline concentration is relatively high, the absolute levels correlating with the difficulty of the birth; it probably helps protect the neonate from hypoglycaemia. Noradrenaline also has a significant part to play in thermoregulation as it causes metabolism of brown adipose tissue (see 2.5.2.1).

2.8.5 *Islets of Langerhans*

Insulin and pancreatic glucagon are both secreted prenatally. Neither, however, seems to be concerned with regulation of fetal plasma glucose concentration. The secretion of insulin in response to a hyperglycaemic stimulus is sluggish and relates more to the duration of the stimulus than to its intensity. It appears that insulin in the fetus is more concerned with enhancing amino acid transfer across cell membranes since increased concentrations of amino acids in the circulation produce a rapid rise in insulin secretion. It is thus more concerned with growth than with glucose homeostasis. Over the first few weeks of extrauterine life, however, the insulin response to hyperglycaemia develops, and although the pancreatic B cell still responds to increases in amino acid concentration, it gradually assumes its adult role.

Less is known about pancreatic glucagon, but again the major stimulus appears to be amino acids and not hypoglycaemia. Hypoxia at birth increases glucagon secretion.

2.8.6 *Thyroid*
Thyroid hormones are found in the fetus from mid-gestation onwards. They are important prenatally for normal development, growth and differentiation of tissues, particularly bone, cartilage and nervous tissue.

At birth, thyrotrophin (TSH) concentration in the plasma rises sharply, possibly because of exposure to a cold environment, and plasma concentrations of thyroxine and triiodothyronine are also raised. This relatively hyperthyroid state is reduced to relative hypothyroidism after about 1 month of extrauterine life and thereafter adult levels are gradually achieved.

2.9 Nervous system

Birth is of little singnificance in the development of the nervous system. Although it is possible to plot growth velocity curves and to present a timetable of myelination, the process is a continuous one from conception to full maturity; birth is merely an incidental milestone along the way. The only ways that birth alters development of the nervous system are:

(1) the alterations in the cardiovascular system allow more blood to flow to the lower limbs and this enhances growth and muscular development, with resulting effects on the nervous control mechanisms;
(2) the sensory systems of the brain are subjected to an increasing number of stimuli.

Readers requiring more details of development are referred to *Scientific Foundations of Paediatrics*; or, for milestones of normal development, *The Normal Child*.

2.10 Further reading

Davis, J.A. & Dobbing, J. (eds) (1981). *Scientific Foundations of Paediatrics*, 2nd edn. Heinemann: London.

Dawes, G.S. (1968). *Foetal and Neonatal Physiology*. Year Book Publishers: Chicago.

Hytten, F.E. & Chamberlain, G. (eds) (1980). *Clinical Physiology in Obstetrics*. Blackwell: Oxford.

Illingworth, R.S. (1979). *The Normal Child*, 7th edn.Churchill Livingstone: Edinburgh.

Langman, J. (1981). *Medical Embryology*, 4th edn.Williams & Wilkins: Baltimore.

Phillip, E.E., Barnes, J. & Newton, M. (1977). *Scientific Foundations of Obstetrics and Gynaecology*, 2nd edn. Heinemann: London.

Young, M., Boyd, R.D.H., Longo, L.D. & Telegdy, G. (1981). *Placental Transfer. Methods and Interpretation*. Saunders: London.

Chapter 3

Old age

Summary

This chapter deals with the physiological changes that occur in old age. The meaning of old age and the life span in humans is reviewed, followed by a discussion of factors that may increase life span. Demographic changes in population are dealt with briefly. There is a section on cellular ageing, which is followed by a discussion of the problems that arise when one tries to investigate the ageing process. This leads into a brief description of some of the theories of ageing, both as they affect the cells and as they affect the body as a whole.

Finally, the changes in the different organs of the body that can be attributed to ageing are detailed. This, of necessity, is a brief survey and students should consult the further reading cited at the end of the chapter for more details.

3.1 Ageing in perspective

Some men (and more women) live for 100 years. Other mammals have a different life span; for example mice rarely live beyond 3–5 years and dogs rarely exceed 20 years of age. In many respects the physiology of the animal as a whole, and the organ systems in particular, is very similar. Why, then, should one species live longer than another? It is also fairly obvious that a man of 80 years has a more wrinkled skin, more grey hair (unless completely bald), poorer vision, poorer hearing and a reduced muscle mass when compared with the same individual at 20 years of age. The cause of these changes is studied by gerontologists. Speculation about the causes of ageing have been rife for hundreds of years, but all attempts at 'cure' have met with a signal lack of success.

3.1.1 *Definitions*

'Ageing' and 'senescence' are terms that are commonly used loosely and interchangeably, although, strictly, senescence in animal biology is the post-reproductive phase of the life span. Ageing implies a decreased viability or an increased vulnerability to external or internal stresses; in other words ageing may be characterized as a decreased ability to maintain homeostasis. As such it is a stage on the pathway towards death.

Not all the changes that occur as an individual grows older can be regarded as ageing changes. In particular, a distinction must be drawn between the debilitating illnesses that many of the elderly suffer and the process of ageing itself, though this may be very difficult. To qualify as an ageing change, four criteria should be met: it should be universal, intrinsic, progressive and deleterious.

Although ageing changes should be universal, this does not imply that all individuals will manifest them to the same degree at the same age. For example, some individuals have completely grey hair by the age of 40 whereas others have only a few grey hairs at the age of 70; nevertheless, grey hair is one manifestation of an ageing change. By intrinsic is meant that the change must be initiated within the organism and not by an external agency; it is the response to external stimuli that is altered. For example, ultraviolet light has different effects on the physical and chemical properties of the proteins of the skin in young and elderly subjects; the change is caused by changes in the skin collagen with age and is not dependent on the ultraviolet

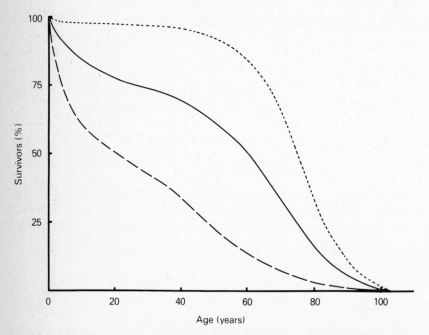

Fig. 3.1 Survival curves for various populations ---- = Britain, 1975; ——— = Britain, 1901; —— — = British India, 1921–30. Survival at earlier ages has increased with passage of time and in more developed countries. Note that the maximum age has not altered.

light. The change also has to be slowly progressive, continuous and irreversible; thus the development of atheroma could be regarded as an ageing process whereas coronary occlusion could not. It is usually assumed that ageing changes are deleterious, although this is the most controversial of the four criteria. For example, increasing crosslinkages between collagen molecules are frequently cited as an ageing change, but 'young' collagen is weak whereas mature collagen is stronger. Is this, then, a deleterious change?

3.1.2 *Length of life*
Death is the one certainly of life. What is not certain is when it will arrive. If one examines survival rates for different populations, the *average* age of death is higher in developed than in less developed countries and has risen over the years (see Figure 3.1). The maximum age, however, tends to be constant at approximately 100–110 years. Advances in medicine have not succeeded in extending the life span but have helped more people more nearly to achieve it.

From census data it is possible to compute the probability that a person of any given age has of dying. Gerontologists are interested in such calculations because age–mortality relationships may show whether ageing changes (rather than pathological changes) are occurring in the individuals of a population. If ageing implies increased vunerability to insults that may result in death (see 3.1.1), there will be a progressive increase in the age-specific death rate (the proportion of people alive at the beginning of the age interval who die during the interval, sometimes expressed as the force of mortality) in the population. This increase was shown many years ago by the English actuary, Gompertz, to be exponential and can be expressed mathematically as:

$$R_t = R_0 e^{\alpha t}$$

where R_t is the mortality at age t, R_0 is the hypothetical mortality at age $= 0$ (actually the age of maturity, which Gompertz took as 35 years) and α is the rate at which the force of mortality increases in the ageing population — sometimes called the Gompertz coefficient (see Figure 3.2).

3.1.3 *Factors that increase life span*
The maximum life span for humans is about 100–110 years but there are only about 1000 people currently living in Britain who have exceeded 100 years. Most of these are women; the average length of life is greater in the female and at 75 years of age they outnumber men by 2 : 1; at 100 years of age this has increased to 6 : 1.

Documentary evidence of longevity is sparse although unsubstantiated anecdotal evidence abounds. There is documentary evidence for an

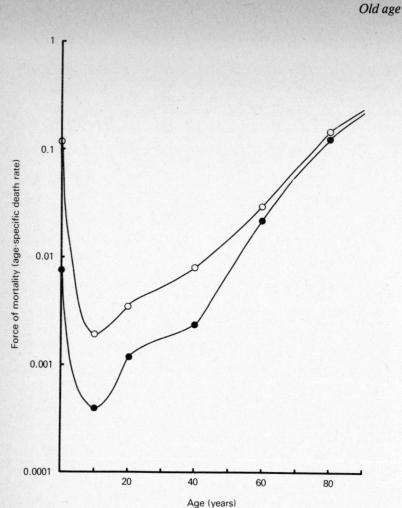

Fig. 3.2   Gompertz plot of force of mortality for English males in 1910 (—○—) and 1960 (—●—). (Based on figures from Registrar General's Decennial Supplement 1961.) Note decreased mortality, especially at earlier ages with the passage of time. Gompertz coefficient (see 3.1.2) applies to ages over 40.

Amerindian in Ecuador reaching 142 years and this area of the Andes seems to support many very old people. Other peoples with an impressive longevity are the Abkhasians of Russia and the Hunzas who live on the borders of Pakistan and China. Major problem exist in that even documentary evidence may have been modified (for example, to escape military service) and the custom of sons adopting their fathers' names makes parish records difficult to verify. All these groups have in common a low density of population, a frugal diet with little animal protein, close-knit family groups and a physically active life associated with an agricultural economy. Whether any of these factors is significant in influencing longevity is not known. It is salutory to reflect, however, that the Amerindians of Ecuador consume on average two to four cups of unrefined rum and 40 to 60 cigarettes per day!

Undoubtedly, many factors that increase longevity are genetic and it is a truism to say that in order to achieve long life one should select one's parents and grandparents with great care.

Several 'environmental' factors have been tested in animals to determine whether they alter life span. Growth retardation produced by dietary caloric deficiency in rats in the immediate post-weaning period has enabled them to survive far beyond the expected life span. Experiments prior to weaning have interfered with development and shortened life span, whereas later in life they show no significant effect. Adrenal steroids have similar effects but the rats cannot leave a sterile environment. Exercise may also have an effect. In the fruit fly, *Drosophila melanogaster*, an increased frequency of sexual intercourse has been shown to have deleterious effects on life span, but, in spite of much speculation, there is no evidence that this finding is applicable to humans. One factor which does seem to be of considerable importance is the age of the mother at conception. Offspring born early in a mother's reproductive life tend to live longer than those born at the end.

3.1.4 *Effects of ageing on population structure*

Animals in the wild rarely age, for whenever an animal begins to 'fail' it is left to die, or may be killed by predators or sometimes even by its own kind. Humans sometimes act differently and as a result the number of older people is growing. It can be seen (Figure 3.1) that in Britain, from the age of about 50 onwards, the number of survivors per 100 000 of an initial population begins to decrease steeply, but whether this signals ageing changes is not known. The World Health Organization regards 45–59 as middle age, 60–74 as elderly, 75–89 as old and 90 + as very old. Perhaps it is better to recognize with Hippocrates that up to 70 is the springtime of old age; 70–75 is the green old age; 75–80 is real old age; 80–90 is ultimate old age; and 90 + senility.

In Britain at present there are more than 9½ million people over 65 years of age. Predictions from demographic trends suggest this will increase. It is estimated that the number of people who are over 60 will increase by 50 per cent by the year 2000, when there will be over 3½ million people over 75. The study of gerontology is thus not only desirable, it is imperative. As the population of this country becomes older, some 20–25 per cent of the population will be showing ageing change. This is likely to pose the greatest problems to the health services in the future.

3.1.5 *Cellular ageing*
So far we have considered ageing as a process applying to the body as a whole. Does ageing occur in individual cells? In the mature body there are basically three types of cells:

(1) highly specialised cells that have lost the ability to reproduce;
(2) highly specialised cells that are still capable of division but usually do not divide frequently;
(3) less specialised cells that readily undergo cell division.

The last two types can be grown in tissue culture.

It was observed more than 60 years ago that cultured chick heart muscle cells divided so many times that the life span of the tissue culture greatly exceeded that of the animal from which the cells were originally derived, and it was postulated that cells might be 'immortal'. However, when great care is taken to prevent transfer of new cells (in the chick serum), such cultures also seem to have a limited life. The number of divisions that occur before death is dependent on the tissue of origin, the life span of the animal and the age of the animal at which the tissue was taken for culture.

There are many exceptions to this rule but, on close examination, all cells which in culture appear to be 'immortal' are abnormal in their chromosomal make-up. For example, the well known HeLa cell line (derived from a cervical carcinoma from Helen Lane) has been cultured for more than 30 years, but each cell now has between 50 and 300 chromosomes. In conclusion, it appears that normal cells, like whole organisms, have a limited life span.

3.1.6 *Studies of ageing*
There are very simple yet cogent reasons why it is difficult to perform ageing studies in man. If one is to avoid dramatic changes in external influences, such as the changing quality of medical care, public health and hygiene, it is necessary to follow a cohort of individuals throughout their lives. This cannot be done by one investigator; a team would need to work for, perhaps, over a hundred years to follow the entire cohort from birth to the grave.

So-called cross-sectional studies are more practical but more difficult to interpret because different populations, which have been subjected to different external influences, are studied at different stages of their life. It must be realized that in such studies members of the older groups are 'selected': they are the fit ones who have endured the rigours of life and thus may not be typical of their cohort.

Animal studies are even more difficult. Senescence is relatively rare in the wild, although some species (albatross, mountain sheep, Arctic fin whales and tortoises) undoubtedly live to great ages. Most survival curves depend

Table 3.1   Life span of fibroblasts.[a]

Species	Number of times population of fibroblasts in culture doubles	Maximum life span (years)
Galapagos tortoise	90–125	175 (?)
Man	40–60	110
Mink	30–34	10
Chicken	15–35	30
Mouse	14–28	3.5

[a] From Hayflick, L. (1975). *Federation Proceedings* **34**, 9–13.

on records from institutions such as zoos and are not necessarily representative of what happens in nature. It is also difficult to extrapolate from these studies to man.

Studies in tissue cultures are much easier to perform, but the relationship between ageing changes *in vivo* and *in vitro* is not yet clear. There does, however, appear to be a crude relationship between the number of times some tissues (usually fibroblasts) will divide in culture and the life of the animal from which they are derived (see Table 3.1). Studies of *in vivo* division potential (based on transplantation and other studies) of haemopoietic and intestinal stem cells show no reduction with the age of the donor.

The problems with studying ageing are reflected in the numerous theories proposed to explain it. These are now discussed.

3.2 Theories of ageing

Confusion about the causes of ageing is reflected by the number of hypotheses (more than 200) currently held to explain the observed facts. Some of these theories refer only to cellular aspects of ageing whereas others are more applicable to the whole organ. Predictably, most excitement is generated about those hypotheses that derive from the fashionable and more general areas of biology, although such hypotheses may have little to do with the real causes of ageing.

Much effort has been expended in trying to reach a unifying theory of ageing. This approach may not be useful since ageing seems likely to be the result of several concurrent changes. The theories that follow have been grouped and are not dealt with in detail.

3.2.1 *Hypotheses of cellular ageing*
These hypotheses assume that there is a time-dependent degradation of cellular information. If this error leads to death of the cell, it can become

very significant if the cell is one of the group that cannot reproduce (see 3.1.5). In spite of dividing the theories into groups, it may yet be that all the factors discussed below interact to produce ageing.

3.2.1.1 *Primary error hypothesis.* This hypothesis postulates that there is some alteration in the information carried by DNA and that this misinformation is transmitted to the next generation of cells. Damage might occur from mutation, from macromolecular damage (such as breaking or cross-linking of the DNA strands), and from some sort of inbuilt, programmed damage that controls alterations in the DNA. All these mechanisms lead to degeneration of the DNA strands. Whatever the cause, the genetic information passed on to the next group of cells is defective; 'abnormal' cells develop and ageing results.

A variant of this hypothesis recognises that breaks in the DNA strands are much commoner than originally thought, and suggests that ageing may be due to a deficiency in the enzyme systems that normally repair the breaks.

3.2.1.2 *Errors in transcription or translation.* DNA in the cell nucleus is used as a template for RNA synthesis (transcription); RNA is then used to produce proteins (translation). Any errors in this complicated process could result in an incorrect amino acid being inserted into a cellular protein. A time at which errors can easily occur is when RNA polymerase is being used to assemble messenger RNA from its constituent bases.

The effect of these abnormal proteins depends on their intended function. If the protein is concerned with structure or metabolism, such as glucose-6-phosphate dehydrogenase which has only a short life in the cell, then the error protein will be quickly eliminated. On the other hand, if an error occurs in a long-lived protein, particularly if it is itself involved in the translation or transcription process, an 'error catastrophe' will result because abnormal proteins will continue to accumulate and the cell will die.

3.2.1.3 *Free radical hypothesis.* Free radicals are chemical species that contain an unpaired electron in an outer orbital; this makes them very reactive. They may be produced transiently during metabolism. Such radicals may attack DNA or proteins, but one of their major effects would be to cause peroxidation of membrane lipids (particularly unsaturated fatty acids), which would interfere with many cellular processes. When free radicals damage other molecules they tend to produce yet more free radicals.

It this hypothesis is tenable, then antioxidants ought to protect the life of cells in tissue culture and perhaps even protect animals. This is true in a number of cases; unfortunately, the design of the experiments allows alternative explanations and so interpretation of the results is always ambiguous.

3.2.1.4 *Programmed ageing.* All the hypotheses just described are based on randomly occurring events. It is difficult to imagine how life span should be relatively fixed for each species, yet variable between the species, if stochastic events are responsible. Proponents of the programmed ageing hypothesis argue that there is some direct genetic control.

Susceptibility to molecular damage varies from one species to another. This may by because of slight genetic differences or because different species vary in their ability to tolerate or repair damage. In this latter respect it must be noted that many copies of genes are present in most species; for example, some genes are repeated five to ten times in bacteria but some 50 to 600 times in vertebrates. The more copies of a gene there are, the less likely it is that random events will have an effect.

3.2.2 *Ageing of the whole organism*

This set of hypotheses, while not denying the theories of cellular ageing, concentrates on changes in whole tissues or changes scattered throughout the body.

3.2.2.1 *Wear, tear and exhaustion.* There have been recurrent suggestions that ageing is due to wearing out of non-replaceable body components. In carnivores, loss of teeth would have this effect; man has circumvented this particular problem. Similarly, however the kidney loses nephrons with age (see 3.3.7); by the age of 70 an average of 40 per cent of the nephrons are lost. As a result, drugs and other harmful substances which are normally excreted on the urine accumulate in the body and may produce toxic side effects.

Steady exhaustion of irreplaceable substances is a variant of the wear-and-tear hypothesis. Such exhaustion probably occurs, but its importance in causing ageing is difficult to assess. Physiological energy reserves are one commodity that it is alleged can be exhausted. However, it may be objected that wear and tear is the result, and not the cause, of the ageing.

3.2.2.2 *Excess accumulation.* Accumulation of substances in inappropriate places is also thought to cause ageing. Abnormal proteins (see 3.2.1.2) accumulate with age but there is evidence of increased protein turnover, suggesting that abnormal molecules are removed. When no abnormal proteins are present, protein turnover is reduced.

Lipofuscin is a pigment that accumulates in the cytoplasm of many different types of cell. The structure and composition of the pigment granules vary from species to species and from tissue to tissue. Lipofuscin may represent increased degradation of defective subcellular organelles. The correlation between pigment granules and altered function is not easy to demonstrate.

In many tissues the amount of calcium is known to increase with age. A notable example is the deposition in the media or the subendothelial layer of large arteries. However, changes in elastin structure may pre-date the changes in calcium accumulation. Such changes may also play a part in the lipid deposition which occurs in arteries with age.

Collagen is laid down between myocardial cells, in the skin and in the lungs as age advances. Such deposits may give rise to many functional changes seen in these tissues. Again, this deposition of collagen may be a consequence rather than the primary cause of ageing.

3.2.2.3 *Endocrine changes*. Hormones are important in homeostasis, and homeostatic mechanisms are disturbed in ageing. Attempts to halt or reverse the ageing process by hormone supplementation have met with a singular lack of success.

Plasma concentrations of thyroxine, corticosterone, growth hormone, adrenocorticotrophin and thyroid-stimulating hormone do not change significantly with age, and those of pituitary gonadotrophins (follicle-stimulating hormone and luteinising hormone) increase with age because the normal negative feedback mechanisms are not operative. These hormones may induce changes in several organs, but there is no indication that endocrine changes are the primary cause of ageing.

Since there are many complicated and important interactions between hormones, it may not be helpful to search for a change in the plasma concentration of a single hormone as the cause of ageing. In addition, a change in the sensitivity of target tissue to circulating hormones could be just as important as a change in hormonal secretion. For example, plasma insulin concentration is not reduced in older people, but the response to insulin is diminished because of altered receptor density.

3.2.2.4 *Cerebral 'clock' hypothesis*. A neuroendocrine clock, possibly situated in the hypothalamus, has been postulated to control ageing processes. Such a 'clock' might be able to control neurotransmitters and hormones, which could then cause homeostatic changes throughout the body. It might even be able to modify transcription and translation processes through hormones.

If such a clock exists, and the evidence for it is meagre, there are a number of exciting possibilities. Perhaps senescence could be influenced by mental processes using biofeedback techniques, thereby validating the adage 'you're only as old as you feel'.

This is probably an entirely different 'clock' from that involved in circadian rhythmicity (see Chapter 6).

3.2.2.5 *Ageing and the immune system.* This hypothesis to some extent bridges the cellular and the 'overall' theories of ageing. Mutation or damage to lymphocytes might cause the manufacture of abnormal antibodies against an individual's own proteins. Alternatively, cells themselves might be altered so that they are no longer recognized as belonging to that individual. In either case there would be an autoimmune disease. It is certainly true that some diseases associated with ageing (e.g. adult-type diabetes mellitus, giant cell arteritis, amyloidosis) are related to immunological abnormalities so that tests that depend on a competent immunological system, like the Mantoux test for tuberculosis, become negative in the elderly. It is also possible that immunological surveillance becomes impaired with age, so that abnormal cells, which would normally be destroyed, no longer are.

3.2.3 *Summary of causes of ageing*
It can be seen from the above list, which is not exhaustive, that there is no consensus on the cause of ageing. Life can be viewed as a series of interacting systems of unstable chemicals; with the passage of time it may be that they just become more disordered. Changes occur simultaneously at many different sites; the theories of ageing are not mutually exclusive. The search for a primary cause of ageing continues.

3.3 Effects of ageing

Changes with age occur in all major organ systems in the body. In some organs, such as brain or muscle, where there is no replacement of dead cells, these changes can be marked and obvious, but in tissues such as intestinal epithelium or red blood cells, where continual replacement occurs, the changes are not usually noticeable. Co-ordinated, complex functions are affected more than simple ones, e.g. reaction times are affected much more than conduction velocity in the nerves concerned. A selection of changes is presented in simplified form in Figure 3.3 to illustrate how widespread the changes are.

Many of the changes (particularly those in the skin, cardiovascular system, lungs and lens of the eye) are due to loss of elastic tissue or, at least, to changes that render the tissues less elastic.

There is a fine distinction between diseases *associated with* old age and changes *due to* old age (see 3.1.1). The changes presented below are ageing changes that occur, to a greater or lesser extent, in all old people.

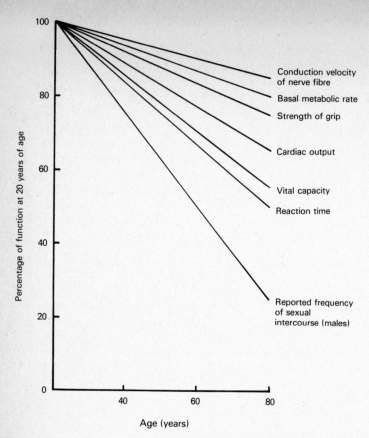

Fig. 3.3 Simplified diagram showing the deterioration of various physiological functions (as a percentage of that function at 20 years of age). Data from Bromley, D.B. (1974), the *Psychology of Human Ageing*, 2nd edn. Pelican Books, p. 93; Copyright © D.B. Bromley 1966, 1974. Reprinted by permission of Peuguin Books Ltd.

3.3.1 *Nervous system*

It is generally stated that brain cells are lost at a rate of about 10 000 per day almost from birth. The evidence for this is not particularly good and is based on cross-sectional studies (see 3.1.6) of post-mortem tissue. As new techniques (e.g. nuclear magnetic resonance (NMR) imaging and computer-aided tomography (CAT) scanning) are applied to ageing research, our knowledge of these processes should increase and allow anatomical changes to be related to changes in function. Some of the changes may be due to an impaired blood supply.

3.3.1.1 *Neuroanatomical changes.* Brain weight may decrease by 6–7 per cent between 20 and 80 years of age. Topographical changes in the various sulci and gyri on the brain surface result in a decrease in surface area of the cerebral cortex. So-called senile plaques occur in the hippocampus and limbic system; these are composed of degenerating neuronal processes, amyloid material and invading neuroglial cells. Neurofibrillary tangles, with the same distribution, consist of dense bundles of thickened neuro-fibrils. Many nerve cells accumulate lipofuscin.

3.3.1.2 *Neurochemical changes*. Depletion of some neurotransmitters occurs with age. There is a decrease of dopamine in the basal ganglia which may produce symptoms of Parkinson's disease; a depletion of noradrenal-ine and 5-hydroxytryptamine in the hypothalamus may be associated with depression; low concentrations of the monoamine metabolite, homovanillic acid, and of acetylcholine in the hypothalamus may be associated with senile dementia. However, it may not yet be valid to assoicate deficiencies in specific neurotransmitters with the various changes of old age.

3.3.1.3 *Higher nervous function*. Little is known about the physiology of higher nervous function; small wonder, then, that it is difficult to explain the changes that occur with ageing. Mental confusion is a very frequent change in the elderly; they may have difficulty in orientating themselves. This can be a result of many different diseases; the underlying change is not clear. Mental confusion, however, is a very real problem.

Short-term memory is impaired in the elderly although long-term memory is surprisingly well preserved. This short-term loss may be linked to a decreased ability to learn in old age. Conceptual skills and reasoning ability begin to decline early but verbal skills are relatively well preserved. This loss of intellectual capacity can to some extent be compensated by drawing on past experience. Many of these symptoms are progressive but occasionally there are 'attacks' when cerebral function is impaired for a short time and then recovers completely; the commonest cause of these is transient cerebral ischaemia.

3.3.1.4 *Sleep*. Older people need less sleep and the 'quality' of the sleep is changed. Electroencephalographic evidence shows that sleep associated with rapid eye movements (REM) is progressively reduced with age. Nocturia (see 3.3.5) may interfere with sleep during the night. On the other hand, elderly people frequently nap during the day.

3.3.1.5 *Regulation*. Later in life many of the body's nervous control mechanisms are impaired. Loss of postural control is often noted in the stooping posture of old people. They are less able to correct quickly and adequately when the centre of gravity is displaced from an equilibrium condition and this, allied to the weakening of muscles, results in a number of falls. Falls are a major problem in elderly people.

Autonomic regulation is also impaired with age. Control of bladder and bowel function is reduced and incontinence occurs. There are a number of other causes of incontinence, but impaired nervous control exacerbates all of them. Autonomic control of blood pressure (3.3.4) and body temperature (3.3.3) are also impaired.

3.3.1.6 *Peripheral nervous system.* The peripheral nervous system is less affected than central mechanisms. Maximum speed of nerve conduction decreases slightly with age (see Figure 3.3) but the speed of simple reflexes does not alter much. Perhaps loss of motoneurones occurs as part of the general loss of nervous tissue (see 3.3.1).

3.3.2 *Special senses*
All old people have less good hearing and vision than younger people. In many this may progress to blindness or deafness, both of which are severely debilitating.

3.3.2.1 *Vision.* With age a lipid infiltration occurs in the cornea and this produces a white border round the limbus called the arcus senilis. It appears to have no functional significance.

The lens of the eye is composed of elastic tissue in an elastic envelope. Accommodation involves changes in lens shape. As elastic tissue ages, it becomes stiffer and this results in loss of accommodative power of the eye — presbyopia. These changes, which are associated with a progressive recession of the near point, become apparent from about 40 years onwards. Visual acuity is also reduced with age but the reason is obscure.

Changes also occur in the retina. Blood vessels grow into the retina from the choroid and these new vessels may become leaky and give rise to haemorrhages and exudates. Normally, pigment epithelial cells remove damaged cells from the retina but this mechanism deteriorates with age and cells that have been shed tend to remain; rods and cones surrounding these shed cells eventually die. Changes in the character of the vitreous humour give rise to small opaque bodies. These appear as black dots that float across the visual field and hence are known as 'floaters'.

Cataracts, glaucoma and frank retinopathy due to diabetes mellitus or hypertension are much commoner in elderly people but probably do not constitute part of the ageing process.

3.3.2.2 *Hearing.* Difficulties with hearing are very common in old age. The causes are multifactorial and include loss of elasticity of the tympanic membrane and the basilar membrane of the cochlea and damage to the sensory hair cells. The commonest problem is presbyacusis, the loss of tones of higher frequency; in some elderly people this is associated with a hypersensitivity to loud noise. Excess wax in the external auditory meatus can give rise to conduction deafness. Tinnitus and impairment of sound localization occur increasingly frequently with age.

3.3.2.3 *Taste and smell.* The number of taste buds is reduced in old age and this, together with a reduction of saliva production, diminishes the sense of taste. Degeneration of neurones in the olfactory bulb is thought to be the cause of an impaired sensation of smell in older people.

3.3.3 *Temperature regulation*

Control of body temperature is less good in elderly people. In part, this is due to impairment of the control mechanisms in the hypothalamus, but failure to appreciate the temperature of the surroundings also plays a part. In a series of patients who had recovered from accidental hypothermia, experiments showed that the responses to cooling (shivering, reduction in hand blood flow and increase in oxygen consumption; see 5.2.4.1) were absent. Social factors, such as poor diet, poor heating of houses and inadequate clothing, together with a general decrease in metabolic processes, contribution to the problem. In a study of old people in London, a deep body temperature of less than 35.5°C occurred in about 10 per cent of those observed; low body temperatures were commonest in the morning.

3.3.4 *Cardiovascular and respiratory systems*

Changes in the cardiovascular system with age are widespread and can be attributed to a number of causes. In general, atherosclerotic changes, which may be detected in early adult life, increase with age and diminish blood flow to most organs; they may be responsible for many symptoms of brain disease, such as confusion and dementia, and for renal changes. Deposits of lipids in the coronary arteries affect the blood supply to the cardiac muscle, and infiltration of the myocardium by collagen (3.2.2.2) decreases the contractility of the heart. Both these changes limit the function of the heart.

Diastolic blood pressure generally increases with age up to about 65 years, and then remains fairly constant. There is a loss of elastic tissue in the aorta, which causes an increase in systolic blood pressure and pulse pressure. Control of blood pressure through the baroreceptor mechanism is impaired and this can cause severe postural hypotension. Many falls in the elderly can be attributed to rising suddenly from a sitting or recumbent postition.

Respiration is disturbed in old age mainly because of loss of elastic tissue in the lungs which makes the work of breathing more difficult. This does not interfere with function at rest, but it does alter the ability to respond to exercise or other stresses.

3.3.5 *Kidney and body fluids*

The kidney is affected by the general changes that occur in the vascular system, and glomerular sclerosis increases with age. This may be so severe that 30–40 per cent of the nephrons are affected by the age of 80. Areas of the kidney may be frankly ischaemic. Generally, the glomerular basement membrane is thickened but it is not known whether its permeability alters.

Glomerular filtration rate, renal plasma flow and concentrating ability may all decrease by as much as 50 per cent in extreme old age, but surprisingly the plasma threshold for renal excretion of glucose appears to be increased, so that old people with diabetes mellitus may fail to exhibit glycosuria. Drugs that are normally excreted in the urine may be retained because of impaired secretory mechanisms. Many old people suffer from nocturia (passing urine at night) and this may be due to impaired concentrating mechanisms during the night.

In spite of these changes, however, there is little or no disturbance of body fluid volume and composition in old age. It is merely the functional reserve of the kidney and the body's ability to cope with extreme conditions that are altered.

3.3.6 *Gastrointestinal tract*

The most obvious change in the gastrointestinal tract with age is the progressive loss of teeth and consequent restructuring of the mandible.

Many old people have difficulties with swallowing. This may be due to ageing changes in the brainstem nuclei which control swallowing, but a reduction in salivary output plays some part. A hiatus hernia (part of the stomach entering the thoracic cavity) is frequent in elderly people (70 per cent in people over 70) but the reason for this, and whether it is an ageing process *per se*, is not known. As age progresses, secretion of gastric juice becomes less and achlorhydria develops with loss of secretion of intrinsic factor.

The villi of the small bowel mucosa become shorter and broader as they age, thereby reducing the area available for absorption. Some degree of malabsorption, particularly of folic acid, vitamin B_{12}, calcium, iron and vitamin D, almost always occurs.

Movements of the colon are diminished and defaecation is particularly severely affected. Constipation may be a major problem especially in the elderly who are immoblie. Taken together with the incontinence that may occur because of diminished autonomic control (see 3.3.1.5), this may indicate why so many elderly people are preoccupied with bowel habit.

3.3.7 *Integument*

Many visible changes occur in the skin and hair of older people. Because of loss of elasticity of the skin, the number of wrinkles increases with age and these are accentuated by loss of subcutaneous fat. The skin becomes thinner and more likely to suffer trauma. As a consequence, multiple bruising (senile purpura) may be caused by minor injuries. A serious problem is the development of pressure sores. These can occur at any age but because of the fragility of the skin, a blood supply impaired by atherosclerosis, and immobility, they are more likely to occur in older people.

Hair loss occurs with age. In males there is a familial type of baldness which is linked to androgen secretion but general thinning of the hair occurs in both sexes. Axillary and public hair is also reduced but, surprisingly, coarse hairs grow in the nostrils and ears.

Hair becomes grey. Although there is much variation in the timing and the degree to which this occurs, all older people have some grey hairs. This is due to the failure of a single enzyme system, tyrosinase, which is responsible for the pigment formation in hair.

3.3.8 *Skeleton*
3.3.8.1 *Bone and joints.* After the age of 40–45, the amount of bone in the body decreases. In long bones both the internal and external diameters increase with age. Calcification of the bone that remains is frequently normal but there is a loss of bone matrix; this is known as osteoporosis. In women after the menopause, lack of oestrogens hastens this process. Immobility may also play a role (see 7.3.5). It is sometimes difficult to determine when this physiological change becomes pathological. Osteomalacia (impaired calcification of the skeleton) is also common in older people; this may result from vitamin D deficiency or lack of exposure to sunlight. These changes result in pain when the bone is stressed and are an important cause of backache in the elderly. They also predispose to fractures.

All joints are subject to wear and tear. This increases with age and is most marked in the weight-bearing joints. Articular cartilage wears thin and ultimately may be destroyed, and bony outgrowths (osteophytes) develop at the edge of the articular surfaces. These may cause pressure on nearby structures or they may break off and further damage the joint. Some degree of osteoarthritis is present in everyone after the age of 50, especially in the spine, but in many cases this gives rise to no symptoms.

3.3.8.2 *Intervertebral discs.* The fibrocartilaginous discs between the vertebrae are composed mainly of water; the water content decreases with age. Whether this is related to damage that occurs in later life is not known, but as age progresses the annulus fibrosus (the hard outer collar) may become cracked, and either protrude itself or allow the nucleus pulposus (the gelatinous centre) to protrude and press on the nerve roots that emerge from the spinal cord; this gives rise to lumbago, sciatica and a host of other back problems.

3.3.8.3 *Muscle*. It is difficult to observe any changes in muscle structure, even in extreme old age. What does occur is a general diminution in the amount of muscle and this is mainly responsible for the fall in lean body mass which is seen in old age. The reduction is partly a result of reduced muscular activity, a kind of disuse atrophy, and partly due to loss of the motoneurones supplying the muscle (see 3.3.1.6), which results in loss of complete motor units. Decreased secretion of anabolic steroids enhances this process, as does the decrease in muscle blood supply which occurs in the elderly.

3.3.9 *Reproductive system*
Changes in the reproductive system are dramatic in the female but more gradual in the male. Failure of the ovarian response after the menopause (usually 40–50 years) leads to a fall in plasma concentrations of oestrogens and progesterone and a rise in those of follicle-stimulating hormone and luteinising hormone. As a consequence, involution of the uterus occurs and there is thinning of the vaginal epithelium and a diminution of the secretion. The vulva atrophies. Mammary and adipose tissue are reduced in the breasts.

In males, there is a reduction in the size of the testis and a thickening of the basement membrane supporting the germinal epithelium with age. However, spermatozoa are still produced well into old age. Although anecdotal stories outnumber facts, there is good documentary evidence that some men retain their sexual potency to their 90th year. Hormone secretion from the testis is reduced, however, and there is considerable loss of libido (see Figure 3.3).

Another significant, but so far unexplained, change in the male is the hypertrophy of the anterior lobe of the prostate. This has a marked effect on bladder function and retention of urine and may account for some of the nocturia in males (see also 3.3.5).

3.4 Further reading

Brocklehurst, J.C. & Hanley, T. (1976). *Geriatric Medicine for Students*. Churchill Livingstone: Edinburgh.

Brocklehurst, J.C. (ed.) (1978). *Textbook of Geriatric Medicine and Gerontology*, 2nd edn. Churchill Livingstone: Edinburgh.

Carter, N. (1980). *Development, Growth and Ageing*. Croom Helm: London.

Comfort, A. (1979). *The Biology of Senescence*, 3rd edn. Churchill Livingstone: Edinburgh.

Davies, I. (1983). *Ageing*. Outline Studies in Biology No. 151. Edward Arnold: London.

Lamb, M.J. (1977). *Biology of Ageing*. Blackie: Glasgow.

Sinclair, D. (1978). *Human Growth after Birth*, 3rd edn. Oxford University Press: London.

Chapter 4

Abnormal pressure

Summary

Man is exposed to a lowered ambient pressure whenever he ascends above sea level. The main problem associated with life at altitude is that of hypoxia. Acclimatization to altitude involves a decrease in the gradient of partial pressures of oxygen which exists between inspired air and systemic capillary blood. As a result of acclimatization, capillary partial pressure of oxygen decreases less than if no acclimatization had taken place. The process of acclimatization involves modification of ventilation, the exchange of oxygen at the lungs, the carriage of oxygen in blood, and the exchange of oxygen at the tissues. As a result of acclimatization, sea-level man can withstand hypoxic conditions at rest without adverse effects, up to an altitude of about 5500 m. The physiological adjustments shown by sea-level man, however, are not as successful in compensating for a hypoxic environment as those shown by the native highlander. As a result, the work capacity of sea-level man at altitude is inferior to his normal work performance at sea level and to that of the native highlander.

An elevated ambient pressure is experienced during diving and by some tunnel workers. The problems assoicated with life in high-pressure environments are due to the direct effects of high pressures on the air-filled cavities of the body, and the requirement to breathe gases at increased pressure. This latter can lead to marked effects on normal physiology, including nitrogen narcosis and oxygen toxicity. In addition, the increased partial pressure of gases, resulting from breathing hyperbaric gas mixtures, leads to an increased amount of gases in solution in the body tissues. If these are not eliminated rapidly on decompression, bubbles of gas form in the body's tissues leading to the symptoms of decompression sickness.

4.1 Changes in ambient pressure

Since air is compressible, the number of molecules of gas per unit volume of atmosphere is greater at sea level than at altitude. Thus barometric pressure decreases exponentially as altitude increases. As an approximation, barometric pressure halves for each 5450 m of ascent. Thus, an estimated 10–20 million native highlanders who live at an altitude above 3000 m, and sea-level natives who go to mountainous regions for reasons of either work or leisure, are exposed to low ambient pressures.

By contrast, exposure to high ambient pressures is limited to relatively few. Such pressures are found, however, whenever one descends below the sea; due to hydrostatic effects, pressure increases by 1 bar for each 10 m of descent in sea water. High ambient pressures are also often experienced by those constructing tunnels, where air pressure is often kept high to prevent the penetration of water into the workings.

The physiological stresses imposed by high and low ambient pressures are quite different. Therefore, in the following account they will be dealt with separately.

4.2 The effects of altitude (low pressure)

Exposure to high altitudes imposes several physiological stresses on the body. For example, ambient temperature is often lower in mountainous regions than at sea level; on average, air temperature falls by 1°C for each 150 m of ascent. Thus ascent to altitude imposes thermoregulatory demands on the body. Furthermore, relative humidity decreases with increasing altitude and this, combined with the higher wind velocities often found in mountainous regions, may lead to increased insensible evaporative water loss and the dangers of dehydration. The greatest problem associated with life at altitude, however, is hypoxia. Although the proportion of the atmosphere which is oxygen is constant at 20.93 per cent, since barometric pressure decreases with increasing altitude (see 4.1), the ambient PO_2 also decreases by the same proportion. As a result, alveolar and arterial PO_2s also fall, resulting in a form of hypoxic hypoxia termed hypobaric hypoxia.

4.2.1 *The effects of acute hypoxia*

When exposure to altitude is rapid (e.g. in ascent to mountainous regions by helicopter or failure of the pressurization system of an aircraft), the immediate hypoxia that results gives rise to a number of adverse effects. On rapid exposure to altitudes in the range 3000–6000 m, the symptoms associated with acute mountain sickness usually appear (see 4.7.1). In addition, at altitudes above 4000 m, there is usually progressive impairment of psychomotor performance with deterioration of sensory acuity, vigilance, judgement, speed of response and manual dexterity. Above 7000 m, the hypoxia has marked cerebral effects, with consciousness being lost within minutes or within seconds at extreme altitudes (Figure 4.1). However, after a person is exposed to altitude for many hours or days, the symptoms of acute mountain sickness gradually disappear, though adverse effects are always seen at altitudes above 6000 m. Furthermore, in 1978 Habeler and

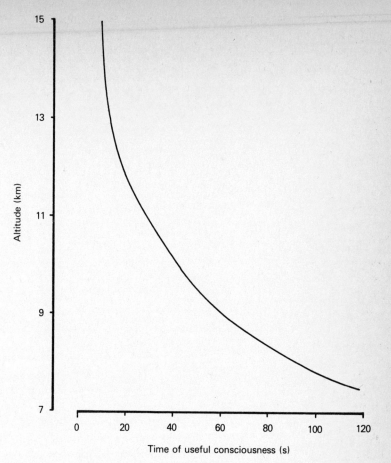

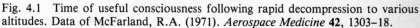

Fig. 4.1 Time of useful consciousness following rapid decompression to various altitudes. Data of McFarland, R.A. (1971). *Aerospace Medicine* **42**, 1303–18.

Messner successfully climbed to the summit of Everest (altitude 8848 m) without the use of supplementary oxygen (or becoming unconscious!). These facts indicate that if exposure to hypoxic conditions is gradual, most adverse effects can be avoided by a gradual acclimatization of the body. The way in which the visitor to altitude and the highland native acclimatize to hypoxic conditions up to 6000 m will now be described.

4.3 Acclimatization to altitude

4.3.1 *Normal oxygen transport*
Normal transport of oxygen from atmosphere to tissues is dependent upon four mechanisms:

(1) pulmonary ventilation — air passing via the trachea and bronchial tree to the alveoli of the lung;
(2) pulmonary diffusion of oxygen — oxygen passing across the alveolar wall and pulmonary capillary endothelium to blood;
(3) transport of oxygen in blood bound to haemoglobin;

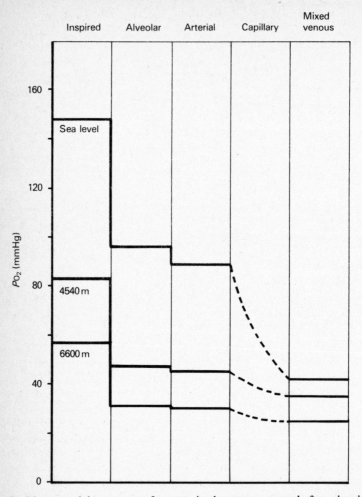

Fig. 4.2 Mean partial pressures of oxygen in the oxygen cascade from inspired air to mixed venous blood. Partial pressures are shown for three groups of subjects: sea-level man, subjects native to 4540 m, and climbers at 6600 m. Data of Luft, U. (1972). In: *Physiological Adaptations. Desert and Mountain*, eds M.K. Yousef, S.M. Horvath & R.W. Bullard, Chapter 10. Academic Press: London, New York; and Hurtado, A. (1964). In: *Handbook of Physiology*, Section 4: *Adaptation to the Environment*, ed. D.B. Dill, Chapter 54. American Physiological Society: Washington.

(4) tissue diffusion — oxygen passing from blood in systemic capillaries to the tissues.

As shown in Figure 4.2, there is a fall in PO_2 at each of these stages of oxygen transport. Thus, oxygen can be thought of as passing down a cascade of pressure from atmosphere to tissues. In healthy man at sea level, the transport of oxygen is efficient and sufficient because of the great partial pressure gradient for oxygen between atmosphere (159 mmHg; 21 kPa) and the cell mitochondria — as low as 1 mmHg (0.15 kPa) in exercise (see 9.2.2.1). At altitude, the fall in barometric pressure (and hence the PO_2 of inspired air) causes a decrease in this pressure gradient. It should be noted that the fall in alveolar PO_2 is proportionally much greater than the fall in ambient PO_2 because:

(a) on entering the trachea, inspired air becomes saturated with water vapour, which exerts a fixed pressure at body temperature;
(b) alveolar PCO_2 remains relatively constant since it is subject to respiratory control.

Acclimatization to altitude involves decreasing the slope of the 'oxygen cascade', as shown in Figure 4.2, by modifying each of the mechanisms involved in oxygen transport as discussed below.

4.3.2 *Changes in pulmonary ventilation*
4.3.2.1 *Changes in the visitor.* The most obvious response seen on initial exposure of sea-level man to altitude is that of hyperventilation. Initially, such hyperventilation is rarely seen at rest until an altitude of 3000 m is reached (corresponding to the point at which alveolar PO_2 (P_AO_2) falls to 65 mmHg (9 kPa)). At altitudes above this, however, hyperventilation progressively increases with increasing altitude, reaching a maximum at 6000 m, where the minute ventilation exceeds that at sea level by 65 per cent (Figure 4.3).

Hyperventilation on initial exposure to altitude is caused by stimulation of the peripheral chemoreceptors of the carotid and aortic bodies, due to the fall in arterial PO_2 (PaO_2). Thus, it disappears if pure oxygen is breathed or if the person is returned to sea level, and it is absent in individuals in whom the peripheral chemoreceptors have been denervated.

The initial hyperventilation has the desired effect of increasing alveolar ventilation and hence P_AO_2. On the other hand, it also leads to more rapid elimination of carbon dioxide, and a fall in $PaCO_2$ results. As predicted by the Henderson–Hasselbalch equation, i.e.

$$pH = pK + \frac{[HCO_3]}{0.03 \times PCO_2 \text{ (mmHg)}}$$

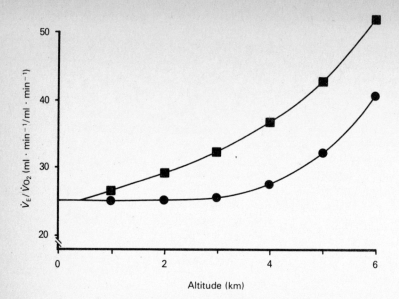

Fig. 4.3 Ventilatory equivalent (minute ventilation/rate of oxygen consumption; $\dot{V}_E/\dot{V}O_2$) at different altitudes in sea-level subjects acutely exposed to altitude (●—●) and after 4 days' acclimatization (■—■). From Lenfant, C. & Sullivan, K. (1971). Reprinted by permission of *New England Journal of Medicine* **294**, 1298–1309.

the fall in $PaCO_2$ leads to a rise in arterial pH and a respiratory alkalosis develops. More importantly, since the blood–brain barrier is permeable to carbon dioxide, the PCO_2 in the cerebrospinal fluid (CSF) falls and its pH rises. It is this increase in pH of the CSF which limits the initial hyperventilatory response. On exposure to altitude, as ventilation rises due to hypoxic drive from the peripheral chemoreceptors, the resulting rise in pH of the CSF tends to inhibit ventilation via its effect on the central medullary chemoreceptors. Until PaO_2 falls to about 60 mmHg (8 kPa) (altitudes up to 3000 m), the inhibition of the central chemoreceptors is sufficient to overcome completely the hypoxic drive from the peripheral chemoreceptors, such that no change in ventilation is seen. At altitudes above 3000 m, as PaO_2 falls below 60 mmHg (8 kPa), the afferent activity from the peripheral chemoreceptors is sufficient to 'overcome' the ventilatory depression from medullary chemoreceptors and ventilation rises, though this ventilatory response is very much less than it would be in the absence of hypocapnia and the fall in CSF hydrogen ion concentration.

Following the initial hyperventilatory response, ventilation in the visitor increases further over the next few days, reaching a maximum after about 4 to 7 days. It can be seen from Figure 4.3 that after 4 days, an increased ventilation is also seen at altitudes below 3000 m, and that at altitudes above 3000 m, ventilation exceeds that seen on acute exposure to altitude, this effect becoming more marked with increasing altitude. Thus, after acclimatization has occurred, the visitor may have a resting ventilation which exceeds that found at sea level by over 100 per cent. Other than at modest altitudes (below 3500 m), this marked hyperventilation is maintained even after many years' exposure to altitude.

This additional increase in ventilation increases P_AO_2 further, thereby reducing the oxygen pressure gradient between environment and alveolar air and increasing the PO_2 to which the pulmonary capillary blood is exposed. This can be seen in Figure 4.2, which shows that the P_AO_2s of men acclimatized to altitudes of 4540 and 6600 m are 47 mmHg (6.4 kPa) and 30 mmHg (4 kPa), respectively; without any ventilatory response the corresponding values would have been 33 mmHg (4.3 kPa) and 7 mmHg (0.9 kPa).

The mechanism of the secondary rise in ventilation seen in the visitor has not been completely resolved. The additional rise in ventilation causes a further fall in $PaCO_2$ which would be expected to worsen the respiratory alkalosis and depress ventilation. Originally it was thought that the secondary rise in ventilation was maintained by renal compensation of the alkalosis, that is, increased renal excretion of bicarbonate lowered arterial bicarbonate concentration and led to a restoration of arterial pH towards normal. However, this renal compensation takes many weeks to develop fully and is too slow to account for the sustained hyperventilation. Subsequently it has been found that the pH of the CSF in visitors at 3800 m is restored to normal (pH 7.32) much more rapidly than is arterial pH by the choroid plexus actively transporting bicarbonate out of the CSF, thereby reducing CSF bicarbonate concentration by 5 mmol·litre^{-1}. Thus, although the PCO_2 of the CSF is decreased on altitude exposure, the fall in bicarbonate concentration will restore the buffer base-to-acid ratio and hence pH towards normal. Since the medullary chemoreceptors are sensitive to local changes in pH rather than PCO_2 directly, the return of CSF pH to normal leads to further stimulation of ventilation by the hypoxia despite a lowered PCO_2. In other words, the medullary chemoreceptors are reset to run at a lower PCO_2. This is confirmed by the observation that once the secondary increase in ventilation has taken place, there is an increased sensitivity to carbon dioxide. This is indicated in Figure 4.4, which shows that the slope of the minute ventilation ($\dot{V}_E$)/P_ACO_2 line is steeper for a given P_AO_2, and the apnoeic point, the intercept on the PCO_2 axis, is shifted to the left after acclimatization to altitude.

More recently the role of changes in CSF pH in the secondary hyperventilation has been challenged since the return of CSF pH to normal has not been confirmed. Rather, a relative alkalinity of the CSF in acclimatized

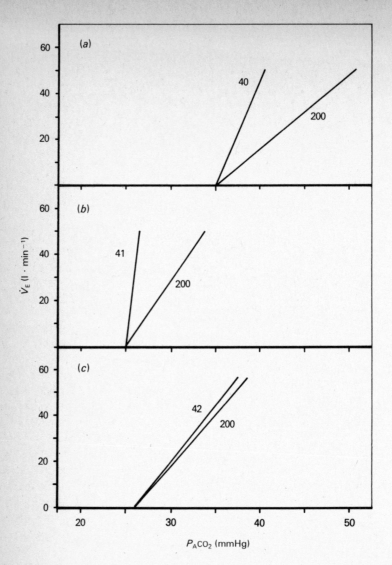

Fig. 4.4 The ventilatory response to varying P_ACO_2 in three groups of subjects: (*a*), subjects at sea level; (*b*) lowland subjects acclimatized to an altitude of 4880 m for 8 weeks; (*c*) native highlanders at 4880 m. The ventilatory responses to varying PCO_2s are shown for P_AO_2 fixed at the figures indicated. Note the steepening of the ventilatory response and its shift to lower PCO_2s during acclimatization to altitude and the lack of change in response when the P_AO_2 was changed in the native highlander. Data of Lahiri, S., Edelman, N.H., Cherniack, N.S. & Fishman, A.P. (1969). *Federation Proceedings* **28**, 1289–95.

visitors has been found. It has been postulated, therefore, that the secondary hyperventilation results from 'central sensitization' of respiratory centres to peripheral chemoreceptor input.

4.3.2.2 *Changes in the native highlander.*

Like the acclimatized visitor, the high-altitude native hyperventilates relative to a normal sea-level man. However, at any given altitude, the ventilation of the acclimatized visitor is greater by about 20 per cent than that of the highland native. Consequently, the PaO_2 of the highland native is lower (and the $PaCO_2$ higher) than of sea-level man acclimatized to the same altitude.

The native highlander differs also by showing a relative insensitivity to hypoxia. Thus the fan of carbon dioxide response curves formed when P_AO_2 is decreased (see Figure 4.4) is very much narrower in the native highlander. This blunted hypoxic response in the high-altitude native persists even after many years at sea level. However, this is not a form of adaptation (i.e. genetically determined), since it does not develop in the offspring of high-altitude ancestry born at sea level. Furthermore, chronic hypoxaemia from birth in sea-level man, such as that found in congenital cyanotic heart disease, also results in a diminished response to hypoxia.

The development of the blunted hypoxic response in the high-altitude native is dependent upon the duration and the degree of hypoxia: it is developed at a younger age and to a greater extent with increasing altitude. The desensitization to hypoxia also develops in adults born at sea level who have spent many years at altitude but the intensity of blunting is less than for the high-altitude natives. The mechanism responsible for the blunted hypoxic response has not been resolved.

The reason for the hyperventilatory response in the highland native in spite of a blunted hypoxic response is also uncertain. However, studies of the CSF in native residents of the Himalayas (4880 m) have shown a pH (7.328) which was closer to that of lowlanders at sea level (pH 7.321) than to that of lowlanders acclimatized to the same altitude (pH 7.374). Similarly, the CSF pH of 16 Andean high-altitude natives (4300 m) was closer to normal sea-level values, in fact slightly acidic (pH 7.295). Thus, it has been proposed that the lower CSF pH acting upon medullary chemoreceptors is sufficient to account for the ventilatory increase found in the highland native in the absence of a functioning hypoxic drive from peripheral chemoreceptors.

4.3.3 *Changes in lung volume and pulmonary diffusing capacity*

Normally, at rest, pulmonary capillary blood reaches diffusion equilibrium with alveolar gas. However, under conditions of hypobaric hypoxia, equilibrium may not be achieved since the steep portion of the oxyhaemoglobin dissociation curve is approached. The increased alveolar ventilation that occurs at altitude limits the fall in P_AO_2 and lessens the reduction in oxygen pressure gradient across the alveolar membrane. Further increases in alveolar gas exchange would be achieved if pulmonary diffusing capacity were increased also. This could be achieved by (a) an increase in alveolar surface area in contact with functioning pulmonary capillaries, and (b) an increase in pulmonary blood volume.

Upon initial exposure of the lowlander to altitude there is a decrease in vital capacity and residual lung volume, but these changes reverse after about 1 month of residence at altitude. Subsequently, measurements of oxygen pulmonary diffusing capacity of adult lowland natives at altitude show no changes when compared with the capacity attained at sea level, even after several years' residence at altitude. In contrast, the highland native has a pulmonary diffusing capacity which is increased by some 20–30 per cent due to changes in both alveolar diffusion characteristics and pulmonary capillary blood volume. As a result the alveolar–arterial PO_2 difference is usually less in native highlanders than in the lowland dweller.

The increased alveolar diffusion characteristics of the native highlander are due primarily to an increased alveolar surface area. Thus, highland natives have a larger lung volume, and especially a larger residual lung volume, than sea-level subjects. Morphometric measurements of the lungs of natives resident at 3840 m have shown alveoli which are larger and greater in number than those in lowland natives of the same body size.

The increased pulmonary capillary blood volume in highland natives is largely due to the increased total blood volume in these people (see 4.3.4.1). In addition, it is likely that the increase in pulmonary arterial pressure seen in highlanders (see 4.6.1), leads to an opening of a larger number of pulmonary capillaries.

4.3.4 *Changes in blood transport of oxygen*
The total quantity of oxygen transported to the tissues of the systemic circulation per unit time is proportional to the product of cardiac output × haemoglobin concentration × saturation of haemoglobin with oxygen. Modification of oxygen transport to the tissues at altitude can involve changes in any one or a combination of these factors, as considered below.

4.3.4.1 *Changes in haemoglobin concentration*. Studies of visitors at altitude and native highlanders have shown a marked increase in haemoglobin concentration over that normally found at sea level. Typically, the haemoglobin concentration in natives at 4540 m is about 12 mmol·litre^{-1} (monomer), thus exceeding the normal by about 30 per cent. This increased haemoglobin concentration results from increased erythropoietic activity. Thus, in this same study, the red cell count was 6.4×10^{12}·litre^{-1} with a

haematocrit of 60 per cent, compared with sea-level control values of $5.1 \times 10^{12} \cdot litre^{-1}$ and 45 per cent respectively, and total blood volume rose from $80\ ml \cdot kg$ body weight^{-1} to $100\ ml \cdot kg$ body weight^{-1}. There is no change in mean corpuscular volume or mean cell haemoglobin content, indicating that the increased haemoglobin concentration results mainly from an increased number of red cells. However, there is also usually a small fall in plasma volume so that some of the increase in haemoglobin concentration can be accounted for by haemoconcentration.

The degree of this polycythaemic response is related to the level of altitude. Up to about 3700 m the haemoglobin concentration rises linearly with increase in altitude. Above this elevation the rise is more rapid with increasing altitude. A large rise in haematocrit, however, limits the effectiveness of the polycythaemic response since the viscosity of blood is increased. This increases the work of the heart and slows blood flow, though oxygen transport is not impaired until haematocrit values exceed about 75 per cent.

The increased erythropoiesis seen at altitude is evoked by the fall in PaO_2, which stimulates the renal secretion of erythropoietin. Erythropoietin stimulates hyperplasia of the erythroid cells in bone marrow. Plasma erythropoietin concentration rises within the first 2 h of altitude exposure but subsequently falls over the next few days to a value intermediate between the normal sea-level concentration and that at the peak response to altitude. This secondary fall is related to the reduction in functional hypoxia as PaO_2 rises due to the hyperventilation and other mechanisms of acclimatization.

Although erythropoietin secretion increases in the first 2 h of altitude exposure, new red blood cells do not appear in the peripheral blood until 3 to 5 days later. Significant increases in haemoglobin concentration and total blood volume are only seen after 2 to 3 weeks, and the full erythropoietic response is not developed until many months at altitude.

The absolute amount of oxygen carried per unit volume of blood does not increase by the same proportion as the haemoglobin concentration since the lower PaO_2 found at altitude leads to a lowering of haemoglobin saturation with oxygen. Nonetheless, at a given altitude, the oxygen content is somewhat greater than the value at sea level. If, however, haemoglobin saturation falls below about 70 per cent, the amount of oxygen carried in arterial blood starts to decline.

From the benefits gained by an increased haemoglobin concentration in blood, it is tempting to think that such compensation would be found in most people at altitude. Although many studies of highland natives indicate a marked polycythaemia, in some the red cell count and haemoglobin concentration are not elevated significantly. Polycythaemia seems to be absent in those native groups which have remained at altitude for many generations and have not ventured to lower altitudes. This possibly indicates that such groups have become adapted rather than acclimatized to hypoxic conditions.

4.3.4.2 *Changes in cardiac output and the systemic circulation.* When the lowlander is first exposed to altitude, there is an abrupt increase in resting cardiac output which is due to an increase in heart rate without alteration of stroke volume (Figure 4.5). This tachycardia results from the hypoxic stress

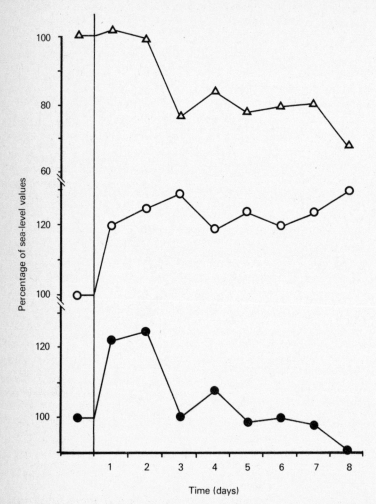

Fig. 4.5   Changes in cardiac output (● — ●), stroke volume (△—△) and heart rate (○—○) in three subjects during the first 8 days' exposure to an altitude of 3800 m (vertical line), expressed as a percentage of sea-level values. Data of Klausen, K. (1966). *Journal of Applied Physiology* **21**, 609–16.

which leads to a generalized increase in sympathetic activity, as evidenced by raised plasma and urinary concentrations of catecholamines. The increased cardiac output persists for no more than a few days, however (Figure 4.5). Thereafter, cardiac output progressively declines so that, after acclimatization has occurred, cardiac output in the long-term visitor is usually found to be comparable with that found at sea level. This secondary decline in cardiac output is primarily the result of a progressive decline in stroke volume with heart rate remaining above sea-level values. The cause of the reduction in stroke volume has not been resolved: possibly hypoxia has a direct depressing action on the myocardium, an effect shown *in vitro* on isolated ventricle.

The resting cardiac output of native highlanders is found to be similar to, or slightly below, that in normal man at sea level. At first sight, this lack of change in both the native and long-term visitor seems unexpected in that a potential method of increasing oxygen transport to the tissues is not being used. However, an increased cardiac output would involve an increased cardiac work and therefore the benefit in oxygen transport would be partially offset by the increased myocardial oxygen consumption.

Though cardiac output does not change at altitude, there is a redistribution of systemic blood flow so that the vital organs receive a larger fraction of the total blood flow than at sea level. As might be predicted, blood flow to the skin is lower in both visitors and highland natives than in lowlanders, particularly when the skin temperature is high. Renal blood flow is also decreased. Despite this decrease, oxygen transport to the kidneys remains normal because of the increased haematocrit and arterial oxygen content. The increased haematocrit and decreased renal blood flow mean that effective renal plasma flow is decreased at altitude. However, renal function remains normal because the fraction of plasma that is filtered increases.

Changes in coronary blood flow during acclimatization are complex and studies are relatively few. Initially, in the visitor there is an increase in coronary flow which is due to the classical vasodilator effect of a low PO_2. After about 10 days, however, coronary flow decreases to about 30 per cent below its normal value. At the same time, oxygen extraction increases to maintain oxygen delivery to the myocardium. In the native highlander, coronary blood flow has also been reported as about 30 per cent less than normal. However, at least in part, this is compensated by the increased vascularization of the myocardium also reported in the native highlander.

4.3.5 *Changes in exchange of oxygen at the tissues*

The final steps in the transport of oxygen to the mitochondria involve the release of oxygen from haemoglobin and its diffusion through the capillary endothelium and tissues. Here too modifications are found at altitude that maintain an adequate flow of oxygen.

4.3.5.1 *Release of oxygen from haemoglobin*. In Section 4.3.4.1 it was shown that acclimatization to altitude involves an increase in haemoglobin concentration. As a result, the oxygen content of arterial blood is maintained or slightly increased despite a lowered PaO_2. Of equal importance is the need for haemoglobin to release sufficient quantities of oxygen at the tissues and maintain an adequate capillary PO_2. Central to this problem is the shape and position of the oxyhaemoglobin (O_2–Hb) dissociation curve (Figure 4.6). An increased unloading of oxygen from haemoglobin can be obtained by a shift of the O_2–Hb curve to the right so that for any given PO_2 the saturation of haemoglobin is less; such a shift might seem advantageous at altitude so that increased amounts of oxygen are liberated from haemoglobin and an adequate capillary PO_2 is thereby maintained.

The position of the O_2–Hb curve has been studied in acclimatized subjects by investigating the P_{50}. (This is the PO_2 giving 50 per cent saturation of haemoglobin at 37°C and pH 7.4 and is usually in the range 26–28 mmHg (3.4–3.7 kPa) in sea-level subjects.) Though the respiratory alkalosis that persists at altitude might be expected to shift the O_2–Hb curve to the left, most studies have shown that in both visitors and native highlanders there is an increase in P_{50} (shift to the right of the O_2–Hb curve). In one study, P_{50} in sea-level subjects rose from 26.6 mmHg (3.6 kPa) to 28.6 mmHg (3.8 kPa) within 12 h of them being transported to 4530 m and rose further to 31 mmHg (4.1 kPa) after 3 days. Highland natives at the same altitude had a P_{50} of 30.7 mmHg (4.1 kPa).

This increase in P_{50} is brought about by an increased concentration of 2,3-diphosphoglycerate (2,3-DPG) within the red cells induced, at least in part, by the respiratory alkalosis. By binding to the β-chains of the haemoglobin molecule, 2,3-DPG favours the stabilization of the deoxy form of haemoglobin so that unloading of oxygen form haemoglobin is facilitated. Thus, in the studies referred to above, the concentration of 2,3-DPG paralleled the changes in P_{50} in the visitors, and in the highland natives 2,3-DPG levels were also high. In addition to its direct effect on the O_2–Hb dissociation curve, 2,3-DPG also increases the Bohr effect (this is the shift of the O_2–Hb curve to the right as pH decreases). The Bohr effect itself is greater in highland natives than in visitors so that an added advantage is gained by the native (Figure 4.6). It is not known how this increased Bohr effect is obtained.

A shift of the O_2–Hb curve to the right allows dissociation of more oxygen at a given tissue PO_2, but it also hinders the loading of blood with oxygen in the lungs. Up to an altitude of about 4500 m, a shift in the O_2–Hb curve to the right seems advantageous since there is little interference with oxygen loading. Above this altitude, however, the deficit in oxygen loading outweighs the advantages gained by the rightward shift of the O_2–Hb curve. In accordance with this, some more-recent studies of high-altitude natives

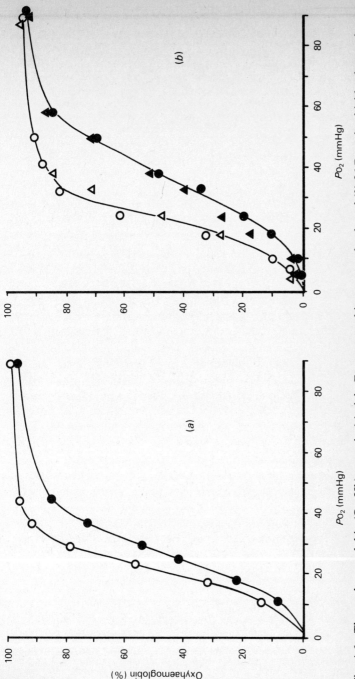

Fig. 4.6 The oxyhaemoglobin (O₂–Hb) curve in (a) eight European subjects at sea level, and (b) 26 Peruvian highlanders native to 4350 m. The curves were determined at pH 6.7 (closed symbols) and 7.4 (open symbols). Note the greater shift (Bohr effect) of the curve with change in pH in the highlanders. From Morpurgo, G., Battaglia, P., Bernini, L., Paolucci, A.M. & Modiano, G. (1970). Reprinted by permission from *Nature* **227**, 387–8. Copyright © 1970, Macmillan Journals Ltd.

have shown P_{50}s that were similar to normal sea-level values. It is thus now believed that shifts in the position of the O_2–Hb curve are of less significance in acclimatization to altitude than was previously thought.

It is interesting to note that man differs from many altitude-adapted animals where a leftward shift of the O_2–Hb curve is often found. In addition, before birth, where the fetus is living in hypoxic conditions, the human fetal haemoglobin dissociation curve is shifted to the left of that for adult haemoglobin (see 2.2.2).

4.3.5.2 *Tissue diffusion*. Oxygen moves from blood through the capillary endothelium and tissues to the cell mitochondria by a process of simple diffusion. By Fick's law of diffusion, the main factors that determine the rate of diffusion are the concentration (or partial pressure) gradient down which oxygen is being transported and the distance over which diffusion is taking place. It has already been described how acclimatization to altitude minimizes the fall in PO_2 and thus the fall in partial pressure gradient between blood and cell mitochondria. Acclimatization also involves minimizing the distance over which diffusion takes place.

Necessarily, most results have been obtained from animals either indigenous to high-altitude areas or chronically exposed to hypoxic conditions; it is only presumed that many of the changes described below apply to man native to altitude.

An increased density of functional capillaries in most tissues but especially the cerebral cortex, myocardium and skeletal muscle has been found in acclimatized animals, though it is not clear if this results from the formation of new capillaries or the opening up of pre-existing ones. Whichever is the case, the net result is the same; diffusion of oxygen is enhanced by a reduction in distance over which it has to take place.

In both acclimatized animals and high-altitude man there is an increased concentration of myoglobin in skeletal muscle, whose function is to act as a reserve of oxygen available for periods of exercise. Furthermore, myoglobin facilitates the diffusion of oxygen into the myoplasm. In some altitude-adapted species there is also an increased mitochondrial density. All in all, these tissue adaptations allow a greater extraction of oxygen from blood.

4.4 Adaptation versus acclimatization

Figure 4.7 summarizes the various mechanisms whereby the lowland and highland native is able to survive chronic hypoxia. It can be seen that many of the mechanisms utilized by the lowlander are qualitatively the same as those utilized by the highland native but, as described in Section 4.3, often they are quantitatively different. This difference might suggest that highland people are genetically adapted to altitude. However, since the

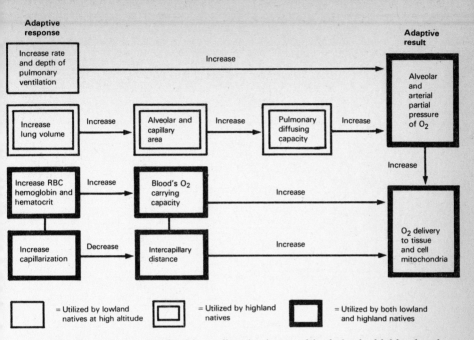

Fig. 4.7 The pathways involved in acclimatization to altitude in the highland and lowland native. From Frisancho, A.R. (1975). *Science* **187**, 313–19. Reprinted by permission, © 1975, the American Assoication for the Advancement of Science.

same responses are often seen in lowlanders who migrate to altitude during childhood, they may develop during childhood growth and development. Thus at present there is little proof of genetic adaptation to altitude in humans. Nonetheless, the greater work capacity of the highland native (see 4.5) would suggest that the natural acclimatization that these people show succeeds in providing a greater compensation for hypoxic conditions than the acclimatization shown by the adult visitor.

It is of interest to note that in many animals which are adapted to altitude, the changes in physiology that allow a normal life in hypoxic conditions are quite different from the acclimatization usually seen in man. In the llama, there is little hyperventilation, polycythaemia or increase in red cell 2,3-DPG concentration. In addition, the O_2–Hb curve is shifted to the left rather than to the right. Such animals compensate for the hypoxic conditions by increasing the extraction of oxygen from blood at the tissues and by more efficient utilization of oxygen, in contrast to the decrease in the 'oxygen cascade' seen in man. Recent observations of certain groups of high-altitude natives in the Himalayas also show little increase in haematocrit and a leftward shift of the O_2–Hb curve, suggesting that they too might be adapted to altitude.

4.5 Work capacity at altitude

When the visitor is exercise-tested on first arrival at altitude, his maximum rate of oxygen consumption per unit body weight, $\dot{V}O_2$max (aerobic capacity), is reduced by up to 50 per cent. Over the next few days, $\dot{V}O_2$max gradually increases as acclimatization takes place but even after several months at altitude it is still lower than that attained at sea level by about 20 per cent.

By contrast, the $\dot{V}O_2$max in the high-altitude native is similar to, or only slightly reduced from, that of sea-level dwellers of similar athletic training. In addition, the native highlander is capable of doing much more strenuous work for a longer time than visitors. This fact has been recognized for many years by successive mountaineering expeditions to the Himalayas where Sherpas are used for portering; often the Sherpanis (Sherpa women) are capable of carrying larger loads for longer periods than the male mountaineers who employ them. As with lung volume and oxygen sensitivity, this increased work capacity does not seem to be genetically determined; those born at sea level who migrate to altitude during childhood attain work capacities, when adult, similar to those of highland natives.

4.6 Other effects of chronic hypoxia

4.6.1 *Pulmonary hypertension*
It is well known that hypoxia induces vasoconstriction in the pulmonary circulation, primarily at the level of the small pulmonary arteries. In the newcomer to altitude this hypoxic vasoconstriction is partly offset by the alkalosis developed on hyperventilation. In the long-term visitor and native highlander, the persistent contraction of the small pulmonary arteries becomes associated with a muscularization and constriction of pulmonary arterioles which, except in the fetal state, are normally devoid of muscle. This leads to a pulmonary hypertension, the degree of which is proportional to the altitude. At rest this pulmonary hypertension is relatively mild; for example, a mean pulmonary arterial pressure of 28 mmHg (3.7 kPa) was found in a study on natives at 4540 m compared with the normal figure of 15 mmHg (1.6 kPa). On exercise, however, the pulmonary hypertension becomes proportionally more pronounced.

In children born at altitude, up to the age of about 5 years the degree of pulmonary hypertension is considerably greater than in the adult. The reason for this is that normally after birth, as arterial oxygenation improves, the pronounced pulmonary vasoconstriction and muscularization of the pulmonary arterial tree found during fetal life (see 2.3.4) regresses. At altitude, however, the chronic hypoxia leads to a much slower rate of regression. In addition, since functional closure of the ductus arteriosus is dependent upon an increased PaO_2 (see Section 2.4.2), the lesser increase in PaO_2 at altitude leads to less powerful constriction of the ductus. As a result there is a high incidence of patent ductus in native highland children, which augments the pulmonary hypertension.

As a result of the increased pulmonary resistance seen at altitude, work of the right heart is increased and right ventricular hypertrophy usually occurs. In addition, the persistent higher pulmonary arterial pressure leads to a thickening of the media of the pulmonary trunk.

4.6.2 *Endocrine changes*
As might be predicted from the increased stress on the body, ascent to altitude is associated with increased secretion of adrenal glucocorticoids and catecholamines. The major increases in plasma concentrations of these hormones are found in the first week of exposure to altitude. Thereafter concentrations fall towards normal.

The increased blood volume found at altitude depresses the secretion of aldosterone and anti-diuretic hormone (ADH). However, the stresses experienced on sudden exposure to high altitude often lead to an oliguria owing to an increased secretion of ADH.

In experimental animals chronically exposed to hypoxic conditions, there is a depression of thyroid function which might be beneficial in increasing the resistance of the myocardium to the effects of hypoxia. The effects of hypoxia on the human thyroid, however, are complicated by the fact that mountainous regions are usually cold and poor in iodide.

4.6.3 *Fertility and pregnancy*
The birth rate in high-altitude regions is much lower than that in lowland regions. In part this is due to sociological reasons but there are biological reasons also.

When sea-level man is transported to altitude, sperm count falls and the incidence of non-mobile and abnormal sperm increases. Only one study has been performed on native highlanders in which normal spermograms were obtained.

The female too shows disturbances of reproductive function on first exposure to altitude. Thus dysmenorrhoea, increased menstrual flow and irregular periods have been reported in women transported to 4330 m. In addition, menarche is delayed on average by 2 years in females born at this altitude.

The fetus lives in conditions which have been described as 'Everest *in utero*' (since PaO_2 is normally about 30 mmHg (4 kPa), equivalent to that in an adult at 8500 m, the height of Everest). At altitude, because of the fall in maternal PaO_2, it might be predicted that there would be a further fall in fetal PaO_2. However, at least in ewes at high altitude, fetal PO_2 is similar to

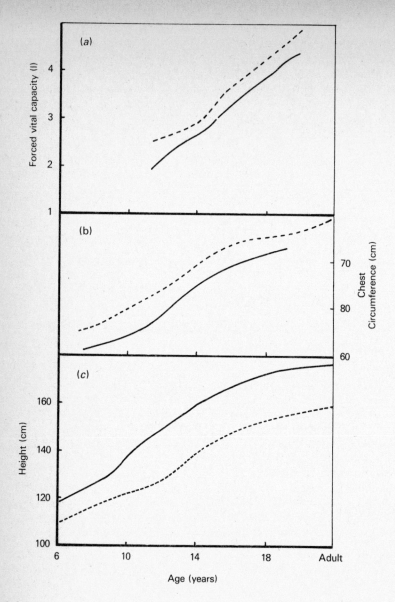

Fig. 4.8 Comparison a (*a*) forced vital capacity, (*b*) chest circumference, and (*c*) stature in sea-level subjects (——) and subjects native to 4300 m (––––). Data of Frisancho, A.R. (1976). In: *Man in the Andes: Multidisciplinary Study of High-altitude Querchua*, eds P.T. Baker & M.A. Little, pp. 180–207. Dowden, Hutchinson and Ross: Stroudsburg.

that found at sea level. The lack of any significant difference in erythro-poiesis between the human neonate at altitude and at sea level indicates that this is probably true of the human also. The lack of fetal hypoxia despite a lowered maternal PaO_2 indicates that there must be increased placental diffusion of oxygen. In accordance with this, the average birth weight at altitude is less than that at sea level but the placenta is of similar size.

4.6.4 *Growth*
The native highlander has a smaller stature than lowlanders of the same age (Figure 4.8). However, the lungs do not participate in this reduction in growth; rather, growth of the lungs is stimulated by hypoxic conditions. The mean chest circumference and vital capacity of the native highlander is greater, therefore, than in lowlanders (Figure 4.8). The advantage to be gained from this, and its effects on pulmonary diffusing capacity, have already been described in Section 4.3.3.

4.7 Mountain diseases

4.7.1 *Acute mountain sickness*
Acclimatization allows the visitor to altitude to lead an essentially normal life, at least at rest, up to an altitude of about 5500 m. However, since acclimatization takes some time to develop, if ascent is too rapid a proportion of visitors, particularly the less physically fit, show a collection of symptoms termed acute mountain sickness or soroche. These symptoms include light-headedness, headache, nausea, lassitude, somnolence, insom-nia, vomiting, anorexia, dyspnoea and muscular weakness. The symptoms usually develop after a time lag of 6–96 h but usually disappear after 3–6 days, though they last longer the higher the altitude and never disappear at altitudes above 6000 m.

Most of the features of acute mountain sickness are due to a redistribu-tion of body fluids. On ascent to altitude, the increased secretion of ADH and adrenal corticoids (see 4.6.2) causes fluid retention. Coupled with the effects of hypoxia, which leads to a decrease in peripheral blood flow and a shunting of blood away from the extremities, there is accumulation of blood in the lungs, brain and splanchnic bed. As a result of this congestion, these vascular beds become mildly oedematous, which explains why many of the symptoms are associated with dysfunction of these organs.

Neither breathing oxygen nor a return to sea level immediately removes the symptoms, showing that they are not due to hypoxia *per se*. There is controversy about the treatment of this condition; most are based upon the removal of fluids by the use of diuretics. However, acute mountain sickness can be best avoided by ensuring a gradual rate of ascent.

4.7.2 *Pulmonary and cerebral oedema*

In some individuals the pulmonary congestion seen in acute mountain sickness develops into frank pulmonary oedema associated with which there is severe dyspnoea. If not treated rapidly by descent to a lower altitude, a fatal outcome is likely. Predisposing factors to pulmonary oedema include a rapid ascent and physical exercise. Thus those most at risk are young men and boys, who tend to be more active. It can also occur in the native highlander who is returning home after some time at sea level. The mechanism of development of pulmonary oedema has not been fully resolved: although pulmonary arterial hypertension is found, pressure in the pulmonary veins and left atrium is normal.

Similarly, cerebral oedema is sometimes more marked, giving rise to severe headaches, or even stupor, paralysis or coma. As cerebral volume increases, CSF pressure rises and venous return is restricted. This commonly causes engorgement of the retinal veins, leading to retinal haemorrhages.

4.7.3 *Chronic mountain sickness*

Chronic mountain sickness or Monge's disease is not a form of prolonged acute mountain sickness. Rather, it is a disease that develops with long periods of chronic hypoxia and is therefore seen mostly in the native highlander. There is no known treatment except to move the patient to a lower altitude.

The symptoms of this disease are akin to loss of acclimatization. Thus, hypoventilation is found, which leads to a further lowering of PaO_2. This, in turn, stimulates further erythropoiesis so that the haematocrit may rise to as high as 80 per cent and haemoglobin concentration to 14 $mmol \cdot litre^{-1}$. Furthermore, the pulmonary arterial hypertension normally seen in the native highlander becomes exaggerated, with mean pulmonary arterial pressures of 50 mmHg (6.7 kPa) being recorded. Why these changes take place has not been elucidated.

4.8 Man in hyperbaric environments

When man dives in the depths of the ocean, he is exposed to a variety of extreme conditions, e.g. cold and lack of light. However, the most marked change is the increase in ambient pressure. Such hyperbaric conditions themselves directly impose several physiological stresses on the body. In addition, air must be supplied at a pressure greater than atmospheric (1 bar). This necessity imposes several other physiological stresses.

In the following sections the problems associated with diving will be considered, but it should be noted that many of the points raised are applicable also to man in a hyperbaric atmosphere, such as that experienced by those constructing tunnels. For convenience, these problems will be considered in two sections:

(a) those due to the direct effects of pressure, and
(b) those due to breathing gases at an increased pressure.

It should be noted from the outset that, unlike man at altitude, there is little evidence of acclimatization to hyperbaric conditions, and since no men live permanently in such conditions there is no adaptation.

4.9 The direct effects of increased pressure

4.9.1 *The problems*

Students familiar with the use of the sphygmomanometer in the determination of blood pressure often wonder why the increased pressures found at depth do not lead to collapse of all of the vasculature. Why this does not occur can best be explained by using an analogy. If a balloon is filled with water and taken underwater, it will not be compressed since the water inside the balloon is incompressible. The external pressure is free to act in all directions on the balloon. Therefore, the pressure of the water inside the balloon rises to equal the ambient pressure and no transmural pressure develops. In the same way, since the body's solids and liquids are virtually incompressible, when the human body is completely surrounded by an increased pressure the tissue pressures rise to the same extent. Thus, for example, there would be no change in blood pressure when measured relative to ambient pressure (though it would be greatly increased when measured relative to atmospheric pressure).

Though these facts are appropriate for the body as a whole, they are not valid for the gas-filled cavities of the body, since gases are compressible. Boyle's law states that at constant temperature the volume of a given mass of gas is inversely proportional to its pressure. Thus, if a balloon filled with air is lowered to a depth of 10 m, where ambient pressure is twice that at sea level, the balloon will be compressed until the volume of the air is half that at sea level. Likewise, unless the mass of gas in an air-filled cavity surrounded by a distensible membrane in the body is increased in proportion to the ambient pressure, a decrease in volume of the cavity will take place so that the internal pressure rises to equal the ambient pressure. However, if the cavity is bounded by a rigid structure, no such change in volume can take place; rather the pressure within the cavity will remain at atmospheric pressure. By analogy, if a hollow steel sphere filled with air to 1 bar is submerged, the pressure within the sphere remains at 1 bar. However, with increasing depth a considerable transmural pressure is developed. When the transmural pressure exceeds the strength of the steel, the sphere will collapse. In the same way, if air is contained in a rigid or semi-rigid cavity in the body and this is not allowed to equilibrate with ambient pressure, the resulting transmural pressure gives rise to the risk of damage to the suspending structures. A more immediate problem, however, results from the effects on the vasculature in any membrane lining such a cavity. Since blood pressure rises (relative to atmospheric pressure) to the same extent as ambient pressure on submersion of the body, if the gas pressure in the cavity is not equalized to ambient pressure, a significant transmural pressure develops across the vasculature. As a result, blood vessels will become distended with blood and there will be increased filtration of fluid out of capillaries. If the increase in transmural pressure is sufficient, blood vessels may even rupture.

The damage resulting from the development of significant transmural pressures is known as barotrauma or 'squeeze'. It should be noted that barotrauma may occur not only on descent but also on ascent. That is, if, during descent, the pressure in an air-filled cavity is equalized with ambient pressure, then, on ascent, as ambient pressure decreases, expansion of the air takes place. If this is not allowed to vent, a significant transmural pressure in the opposite direction will develop.

4.9.2 Barotrauma of the ear and paranasal sinuses
Normally air pressure in the middle ear cavity is the same as atmospheric pressure. Such equalization of pressure is maintained via the Eustachian tube. However, if the Eustachian tube is oedematous, as during upper respiratory tract infection or in an allergic reaction, such equalization may not be possible. In consequence, as a descent continues, a significant pressure gradient develops across the tympanic membrane causing it to stretch and bulge into the middle ear and hence cause pain. In addition, dilatation and eventual rupture of small blood vessels in both the tympanic membrane and the lining of the middle ear occur. If the descent is sufficiently great, the elastic limit of the tympanic membrane may be reached causing it to rupture. This may take place when the pressure difference across the tympanic membrane is as little as 100 mmHg (13 kPa), which occurs at a depth of only 3 m.

As in the case of the middle ear, if the channels connecting the paranasal sinuses to the naso-pharynx are blocked, as in sinusitis, air cannot pass in or out of the sinuses. Thus, on descent, equalization of pressure between the sinuses and the nasal passages will not occur, leading to transudation and haemorrhage of the lining of the sinuses accompanied by pain.

4.9.3 Pulmonary barotrauma
In contrast to the sinuses and middle ear, which are damaged usually on descent, the lungs are more commonly damaged on ascent.

Other than in dives achieved by breath-holding, it is necessary to breathe a gas mixture at a pressure equal to ambient pressure. Otherwise the volume of air in the lungs and airways would change in accordance with Boyle's law. So long as normal breathing is continued and gas is supplied at a pressure equal to ambient pressure, no problems arise in pulmonary function, during either descent or ascent. If, however, after breathing gases at increased pressure at depth, the breath is held during ascent, the volume of gases in the lungs increases and may exceed the normal capacity of the lungs, which ultimately rupture. Such 'burst lung' usually occurs when intra-alveolar pressure exceeds ambient pressure by about 80 mmHg (11 kPa); this may occur after dives to as little as 2 m.

When the lung tissue tears during lung rupture, air can escape from the alveoli and travel to several sites:

(1) into the interstitial spaces of the lung;
(2) into the pleural cavity, resulting in pneumothorax;
(3) along the vascular sheaths to the hilum of the lung; and
(4) into the pulmonary circulation whence it reaches the systemic circulation via the left heart where it may cause air embolism at any site, occluding circulation beyond that point.

Since the diver is usually upright during ascent, the air emboli often enter the carotid artery and lodge in the cerebral circulation. In this case, stroke-like symptoms appear and often consciousness is rapidly lost.

Although there are no problems associated with descent when breathing compressed air, damage may occur during a breath-holding dive when the volume of air in the lungs decreases as depth increases. No difficulty arises until the lung gases are compressed to a volume the same as residual volume. Further increase in depth beyond that point results in the development of a negative intra-alveolar pressure leading to 'lung squeeze' — pulmonary congestion, oedema and haemorrhage.

4.9.4 *Barotrauma at other sites*
Barotrauma can develop in any air-filled cavity in the body that does not equilibrate with ambient pressure. It is thus often found in carious or filled teeth. Barotrauma of the gastrointestinal tract rarely occurs during descent since the wall of the gut is non-rigid so that any gas contained in the gut can shrink in volume. However, if extra gas is swallowed or produced during a dive, then on ascent this will expand causing abdominal distension, discomfort and flatulence.

When a diving helmet is used during diving, a further danger is that of classical 'diver's squeeze'. This occurs when the air supply to the helmet becomes inadequate, perhaps as a result of some mechanical failure in a pump. As the air pressure in the helmet falls, the helmet itself becomes a non-equalized space and the external water pressure tends to force the whole of the diver into it.

4.9.5 *High-pressure neurological syndrome*

Although the body tissues can be considered as virtually incompressible, in very deep dives with large ambient pressures the tissues may be compressed slightly. In particular, lipids are more compressible than water. Thus at extreme depths the lipids in cell membranes are compressed. As a result the normal ionic permeability characteristics of nerves are disrupted and symptoms associated with neurological disruption are seen. These symptoms are termed high-pressure neurological syndrome and include tremors, decreased manual dexterity, dizziness, loss of attentiveness and nausea.

4.10 The effects of breathing hyperbaric gases

When air is breathed at normal atmospheric pressure (1 bar) through a snorkel, a dive below a depth of only 0.5 m is unsafe. This is because the increase in ambient pressure which occurs with increasing depth tends to compress the air in the lungs. To prevent the diminution in lung volume, the inspiratory muscles must exert a greater pressure (equal to the increase in ambient pressure). The greatest pressure that can be generated by the inspiratory muscles is about 90 mmHg (12 kPa): when ambient pressure exceeds this value (which is found at a depth of 1.2 m), inspiration cannot occur. In practice the maximum depth of breathing air at 1 bar that can be safely maintained is much less than this for two reasons:

(1) when alveolar pressure is much less than ambient pressure 'lung squeeze' may occur;
(2) a snorkel above 0.5 m in length adds a considerable deadspace and alveolar ventilation will become inadequate.

In dives deeper than 0.5 m, therefore, air at ambient pressure must be breathed. In most 'scuba' diving for leisure purposes, the diver inspires from a cylinder of compressed air (21 per cent O_2 : 79 per cent N_2). As the diver goes deeper, a demand valve ensures that air is provided at a pressure to match ambient pressure so that normal breathing can continue and no significant transthoracic pressure develops. However, breathing air at pressure imposes two problems which limit the depth to which the diver can go:

(1) as air pressure increases, the densities of the gases increase by the same proportion;
(2) in accordance with Dalton's law, with an increase in total pressure of a mixture of gases the partial pressure of each gas in the mixture increases by the same proportion.

4.10.1 *The effects of increased air density*

As the density of air breathed increases so does the mass of gas moved with each breath and the work of breathing. The maximum voluntary ventilation is approximately proportional to the reciprocal of the square root of the density. This means that at a depth of 30 m (4 bar) the maximum voluntary ventilation of a man breathing compressed air is only 50 per cent of that at sea level. However, in practice, compressed air is rarely breathed for deep dives because of the intervention of nitrogen narcosis (see 4.10.2.1). Rather, the nitrogen in the gas supply is replaced by helium, which has only one-seventh the density of nitrogen so that the same increase in the work of breathing is not found.

Breathing compressed air also increases the work of breathing in another way: as the density of air breathed increases, so flow of air in the airways becomes more turbulent, resulting in an increase in airway resistance. In addition, an increased density of gases hinders their intra-alveolar diffusion.

As a result of these factors, whereas maximum work capacity at sea level is normally limited by cardiovascular transport of oxygen, the limitations underwater are largely ventilatory.

4.10.2 *The effects of increased partial pressure of gases*

As depth increases, the partial pressures of gases in the inspired air increase. As a result, the partial pressure of these gases in arterial blood and tissues increases and therefore, according to Henry's law ('at a given temperature, the mass of a gas dissolved in a given volume of solvent is proportional to the pressure of the gas with which it is in equilibrium'), the amounts of gases dissolved in the body tissues are also increased. This effect is greater in those tissues, such as fat, which have a greater solubility coefficient.

Though both nitrogen and oxygen at the partial pressures found at sea level have no known harmful effects, they do have detrimental effects at the concentrations present in solution when breathing air under pressure.

4.10.2.1 *Nitrogen narcosis.*

Nitrogen acts progressively as an anaesthetic as its partial pressure increases. When breathing compressed air, most divers begin to show the symptoms of nitrogen narcosis at 30 m (4 bar), but some divers show signs at as little as 10 m. Initially, the symptoms resemble those of alcohol intoxication: judgement, thought, and the ability to do tasks that require mental or fine motor skill are all impaired. At depths beyond 50 m, most inexperienced divers can do no useful work though this limit is extended to 75 m for the 'seasoned' diver. At depths beyond 90 m, in most divers unconsciousness approaches, though there are also the added complications of oxygen toxicity (see 4.10.2.2).

The mechanism of nitrogen narcosis seems to be similar to that of anaesthetic gases. That is, being more soluble in lipids, nitrogen causes a small but significant increase in membrane volume. This leads to a blockage of ion channels and consequent disruption of the excitability of the axon and impairment of synaptic transmission. These effects are similar to those seen during high-pressure neurological syndrome (see 4.9.5), where there is compression of the cell membrane, which suggests that there is a 'critical volume' for normal cell membrane function.

Because of these effects, nitrogen in inspired gases is replaced by helium in sustained deep dives since this gas has only 12 per cent of the anaesthetic properties of nitrogen and about half the solubility of nitrogen.

4.10.2.2 *Oxygen toxicity*.

An increased PO_2 is also toxic, having major pulmonary and neurological effects. Initially, the symptoms of pulmonary oxygen toxicity are dyspnoea, pulmonary oedema and intra-alveolar haemorrhage. These symptoms are caused by progressive destruction of alveolar endothelial and epithelial cells. Pulmonary oxygen toxicity can develop when breathing a PO_2 as little as 375 mmHg (50 kPa), a PO_2 found at a depth of 16 m when compressed air is breathed. However, the symptoms arise only after many hours, though the time becomes shorter as depth is increased; thus, they take over a day to develop when breathing a PO_2 of 375 mmHg (50 kPa) but only 5 h when breathing a PO_2 of 1500 mmHg (200 kPa).

Since pulmonary effects are seen only on chronic exposure to hyperoxic conditions, a more common danger in diving is that of neurotoxicity which develops with a shorter latency. The signs of oxygen toxicity on the central nervous system are similar to those of *grand mal* epilepsy. These symptoms are sometimes preceded by vertigo, restlessness, nausea, disturbances of vision, localized muscle twitching and paraesthesia. On average, these symptoms become apparent at PO_2s exceeding 1500 mmHg (200 kPa), which will be reached at 10 m if pure oxygen is breathed. The development of symptoms follows a latent period which shortens with increase in PO_2 and with exertion.

The symptoms are brought about by an increased permeability of neuronal membranes to sodium and potassium, though the mechanism by which these permeability changes are brought about has not been resolved. The concentration of gamma-aminobutyric acid (GABA) in the brain is known to decrease before the convulsive episode and administration of GABA before exposure to hyperoxic conditions is protective. However, other neurotransmitters may be involved. Further, it has been proposed that the cell membranes might be damaged through oxidation of lipids and sulphydryl groups of proteins.

4.11 Decompression sickness

4.11.1 *Gas uptake during compression*

The volume of nitrogen contained in solution in the body is usually about

1 litre at sea level. This volume increases by a further litre for each 10 m descent (each 1 bar increase in pressure) when compressed air is breathed. However, the new equilibrium point is not attained immediately. This is because blood volume is only 8 per cent of total body volume and so considerable time is taken for enough nitrogen to be transported to the tissues for them to come to equilibrium at the new PN_2. Furthermore, the rate of gas uptake and the quantity of gas taken up by any particular tissue are dependent upon the rate of perfusion and solubility coefficient of the tissue. Thus, for example, well perfused tissues, such as the brain, take up nitrogen rapidly whereas adipose tissue, which has a lower perfusion and greater solubility to nitrogen, reaches equilibrium much more slowly.

It is evident that when a descent is made and the partial pressure of gases in the tissues comes into equilibrium with the new ambient pressure, the quantity of gases contained in solution in the body increases in direct proportion to the depth. The problem for the diver is to eliminate these gases when decompression takes place on ascent.

4.11.2 *Gas elimination during decompression*
When the diver ascends and inspired gas pressure decreases, alveolar partial pressures of gases fall and a partial pressure gradient for gases develops between the tissues and alveoli. Therefore, the gases will diffuse from the tissues to blood and be eliminated by the lungs. However, as with gas uptake, this is a relatively slow process. Therefore, if the quantity of gases dissolved in the tissues is large and the rate of ascent is rapid, so that ambient and tissue pressures fall rapidly, the total pressure of the gases in tissues will rise above tissue hydrostatic pressure, causing the gases to come out of solution and form bubbles. (An analogous situation occurs when a bottle of champagne is opened.) The symptoms that result are known as decompression sickness. (This is often referred to as 'the bends' by the diver, though in fact 'the bends' are only one manifestation of decompression sickness.) Of particular importance in the development of decompression sickness is the formation of bubbles of inert gases which cannot be utilized by tissues, particularly nitrogen, when compressed air has been breathed. If oxygen bubbles form, these can be used locally in tissue metabolism and carbon dioxide bubbles do not form since there is little of this gas in inspired air.

4.11.3 *Development of decompression sickness*
A rapid dive to great depth, followed by rapid ascent, may not lead to decompression sickness since little time is available for the tissues to equilibrate with the increased partial pressure of gases. By contrast, a dive to a modest depth of long duration must be followed by a slow rate of decompression since sufficient time is available for the tissues to become saturated with inert gases. For this reason, further advantages to be gained from

breathing an oxygen-helium mixture, rather than compressed air, are that only about half as much helium dissolves in the body as nitrogen and, because of its smaller atomic size, helium diffuses through the tissues 2.5 times more rapidly than nitrogen and therefore can be eliminated more rapidly during decompression.

Additional factors that predispose to decompression sickness are obesity, poor physical condition, age and exertion. Professional divers who frequently undergo decompression are less likely to show any symptoms, suggesting that there might be some acclimatization, but the mechanism of this has not been resolved; it may be associated with an elevated threshold to pain.

When the rate of ascent of the diver is too rapid, bubbles of gas can form at any site within the body; they may be formed intravascularly, blocking capillaries, or extravascularly, distorting the tissues and causing pain. The site of formation of bubbles determines the symptoms of decompression sickness seen, and the number and size of bubbles formed determine its severity. Thus, symptoms can vary from minor localized pain to major lesions in the lungs and central nervous system leading to death. Symptoms may appear within minutes of decompression or occasionally be delayed by up to 24 h. In general, severe symptoms tend to appear early, within the first hour of decompression.

In addition to their direct mechanical effects, the bubbles formed during rapid decompression also produce a number of indirect effects in blood as a result of the gas–blood interface, with consequent disruption of membrane protein structure.

4.11.4 *Symptoms of decompression sickness*
One of the most common symptoms of decompression sickness is that of pain in the joints of the extremities; this is due to the formation of bubbles in the tendons and ligaments around joints. The diver refers to these symptoms as 'bends' when the pain is severe, or 'niggles' when it is mild. In the diver the pain is usually felt in the shoulder and elbow joints, whereas in the tunnel worker it usually appears in the hip and knee joints.

A further problem found in those who have to undergo frequent decompression is that of necrosis of bone, which may in time cause joint deformity or even collapse. The reason for this necrosis is thought to be the formation of minute bubbles with platelet aggregation (see below) which block the end arteries of bone.

Although found less frequently, bubble formation in the spinal cord or central nervous system has more serious consequences. Bubble formation in the venous vertebral plexus leads to obstruction of blood flow to the spinal cord and results in degeneration of nerve fibres. This, in turn, may lead to paralysis which often affects the body below the waist.

Bubble formation in the brain may cause hemiplegia, convulsions, visual disturbances or unconsciousness. In addition, there may be vertigo (called 'staggers' by divers) due to the formation of bubbles in the vestibular apparatus.

Any bubbles formed intravascularly are carried to the pulmonary circulation where they lodge, blocking pulmonary capillary beds. This results in a number of symptoms which divers call 'the chokes'. These include dyspnoea, coughing, substernal pain and possibly cyanosis. If not rapidly treated, this condition may lead to circulatory collapse and death. In addition, the formation of intravascular bubbles and the consequent formation of a gas–blood interface lead to a number of changes in blood. These include clumping of red cells, the adherence of platelets to the bubbles, and activation of the clotting pathway, of the complement system, of kinins and of other smooth muscle activating factors, causing constriction of blood vessels.

4.11.5 *Avoidance of decompression sickness*
The avoidance of decompression sickness is based upon selecting a rate of ascent that precludes significant bubble formation. Most decompression tables are based upon an observation made in 1908 by Haldane and his colleagues that decompression sickness was never found in men who surfaced from depths (up to 10 m) where the pressure was up to 2 bar. From this they suggested that tissues could withstand a decrease in pressure if total tissue gas pressure did not exceed twice the ambient pressure. Thus decompression tables are based upon making an ascent in several stages, making 'stops' at depths where the ambient pressure is half that at the depth of the previous stop. At each stop sufficient time is allowed for the inert gases to be eliminated from the body until equilibrium is again attained.

4.12 Further reading

4.12.1 *Altitude*

Baker, P.T. (ed.) (1978). *The Biology of High-altitude Peoples*. Cambridge University Press: Cambridge.

Frisancho, A.R. (1975). Functional adaptation to high-altitude hypoxia. *Science* **187**, 313–19.

Heath, D. & Williams, D.R. (1981). *Man at High Altitude: the Pathophysiology of Acclimatization and Adaptation*, 2nd edn. Churchill Livingstone: Edinburgh.

Lahiri, S. (1977). Physiological responses and adaptations to high altitude. In: *Environmental Physiology* II. International Review of Physiology Vol. 15, ed. D. Robertshaw, Chapter 7. University Park Press: Baltimore.

Lenfant, C. & Sullivan, K. (1971). Adaptation to altitude. *New England Journal of Medicine* **284**, 1298–1309.

4.12.2 *Diving*

Bennett, P.B. & Elliott, D.H. (eds) (1975). *The Physiology and Medicine of Diving and Compressed Air Work*. Bailliere Tindall: London.

Elliott, D.H. (1981). Underwater physiology. In: *The Principles and Practice of Human Physiology*, eds O.G. Edholm & J.S. Weiner, Chapter 6. Academic Press: London, New York.

Miles, S. & Mackay, D.E. (1976). *Underwater Medicine*. Adlard Coles: London.

Strauss, R.H. (1979). Diving medicine. *American Review of Respiratory Disorders* **119**, 1001–23.

Chapter 5

Altered temperature

Summary

Natural human populations encounter environmental temperatures ranging from
− 50 to + 50°C. Nevertheless, deep body temperature is normally kept within the
narrow range of 36–38°C. This chapter considers how such homeothermy is
achieved despite these thermal extremes.

Responses to cold fall into two major categories, some attempting to increase heat
production, others seeking to decrease heat loss. The former include an increased
muscular activity (primarily shivering) and an enhanced non-shivering thermogenesis, which is of particular importance in the neonate. The latter include
behavioural responses, although the major protection against heat loss comes from
the insulative value of clothes and housing. The major physiological response in this
category involves a cutaneous vasoconstriction. Hypothermia and cold injury are
two of the special problems of cold stress which can develop if the normal
thermoregulatory mechanisms fail to cope with reduced ambient temperatures.

Responses to high temperatures seek to decrease heat gain and increase heat loss.
Decreased activity and reduced feeding lower endogenous heat production; behavioural responses and the use of suitable clothing and housing reduce heat gain from
the environment. Sweating is man's major means of effecting an increased heat loss
in a hot environment. It can also be used to demonstrate the phenomenon
of thermal acclimatization, which is more evident in the heat than the cold.
Cardiovascular responses, including a cutaneous vasodilatation, also serve to
increase heat losses as ambient temperatures rise. The chapter also deals with the
special problems of heat stress, considering heat illnesses having both general and
localized effects.

5.1 Introduction

Man is homeothermic; he normally maintains his deep body temperature
between 36 and 38°C despite wide variations in his own metabolic activity
and in the temperature of his environment. In order to achieve homeothermy, heat production by metabolism must balance heat loss to the
environment. The heat balance equation, $M = E \pm C \pm R \pm Cd \pm S$
relates metabolic heat production (M) to the components of environmental
heat exchange. Here, E = evaporation (of sweat); C = convection; R =
radiation; Cd = conduction; and S = heat stored within the body. Although
E, C, R and Cd usually represent heat losses from the body, the direction of
heat exchange will ultimately depend upon environmental temperature. If
the external temperature exceeds skin temperature, for example, heat
exchange via C, R and Cd may be *from* the environment *to* the body, and
therefore constitute heat gains.

5.1.1 *The thermoneutral zone*

At low environmental temperatures, man maintains his core temperature primarily by increasing his rate of heat production. At high external temperatures, heat losses are of prime importance. The relationship between the components of heat balance and environmental temperature in man is shown in Figure 5.1. In this figure, the 'thermoneutral zone', TNZ, describes the environmental temperature range over which metabolic rate is kept at a minimum. At lower ambient temperatures, metabolic rate is raised to increase heat production and offset heat losses. The point at which metabolic rate begins to rise is called the 'critical temperature'. This varies according to body size and insulation. For a resting, 70 kg, naked man (see Figure 5.1), the critical temperature is around 27°C; for the Arctic fox, with

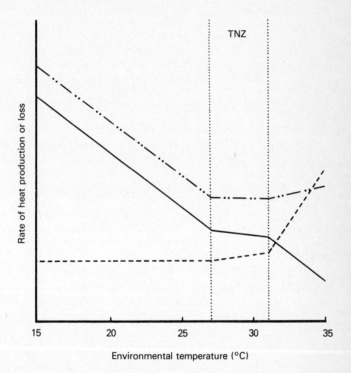

Environmental temperature (°C)

Fig. 5.1 Heat balance at different environmental temperatures in man. Symbols —··—··, metabolic heat production; ——, non-evaporative heat loss; ————, evaporative heat loss; TNZ, the 'thermoneutral zone' (the environmental temperature range over which metabolic rate is kept constant). 'Critical temperature' is the ambient temperature at which metabolic rate starts to rise, that is, the lowest temperature within the TNZ (27° in this diagram). Values apply to a naked, 70 kg male; factors such as clothing, posture, humidity and air velocity will alter the absolute values shown.

its thick winter coat, it will be below $-50°C$. Thermoregulation below the critical temperature is thus reliant on metabolic heat production, where heat is generated during the oxidation of body fuels. Some heat is produced directly in the course of the metabolic oxidation, but some of the energy is conserved first as ATP, whose subsequent hydrolysis yields heat.

At the upper end of the thermoneutral zone (around $31°C$ in man), evaporative heat loss mechanisms are used to balance the combined rates of heat gain from the environment and from metabolic processes. Sweating is central to the heat loss mechanism in man, and the ambient temperature at which it begins is called the 'set point' for sweating. The increase in metabolic rate above the thermoneutral zone (see Figure 5.1) reflects the onset of heat loss mechanisms (sweating included) which attempt to combat the rise in environmental temperature. There are, of course, limits to physiological control. As the environmental temperature falls below the thermoneutral zone, heat loss may ultimately exceed heat production and body temperature will fall. The clinical definition of hypothermia in man is a core temperature of $35°C$ or below. Conversely, as ambient temperature rises, heat gain may ultimately exceed heat loss. Hyperthermia develops at a core temperature above $40°C$.

Physiological responses to cold and heat therefore differ in several respects and it is convenient to deal with the two separately. One factor common to both, however, is the phenomenon of 'acclimatization'. This is a term used to describe physiological changes induced by repeated exposure of an individual to altered environmental conditions. When a *single* environmental factor (e.g. temperature) is experimentally changed, the term 'acclimation' is used. Both terms should be distinguished from 'adaptation', which strictly describes a long-term evolutionary change.

5.2 Responses to low temperature

The geographical distribution of man reveals natural populations in northeast areas of the USSR that regularly experience January temperatures of $-50°C$. Moreover, since the critical temperature of naked man is around $27°C$, most of the Earth's natural environments will present a potential cold stress at some time of the day or year. Man's responses to low temperatures are primarily behavioural. He will avoid exposure to cold, and make use of suitable clothing to improve insulation. Physiological responses include an increased muscular activity and a cutaneous vasoconstriction. There is limited evidence for true cold acclimatization.

Cold responses fall into three main categories: increased heat production, decreased heat loss, and reduced body temperature. The latter, although characteristic of hibernating mammals, is rare in man.

5.2.1 *Increased heat production*

5.2.1.1 *Increased activity.* Increased voluntary muscular activity increases heat production by stimulating ATP hydrolysis on contraction of the muscle fibres (see Chapter 9.2.1.4). Hence, in the cold, man will stamp his feet, rub his hands, and generally move about more quickly. Shivering is a specialized form of muscular activity, and in man is probably the most important physiological mechanism which initially compensates for heat loss at low ambient temperatures. It is a predominantly involuntary process controlled by the autonomic nervous system, whereby skeletal muscle fibres contract and relax periodically. The term 'shivering' encompasses all grades of contraction from an increased muscle tone (pre-shivering tone), via a barely perceptible tremor, to vigorous overt shivering. With pre-shivering tone, muscle contraction is uncoordinated, although the process becomes more regular as cold exposure continues.

Shivering is first apparent in the extensor muscles and proximal muscles of the upper limbs and trunk rather than in the extremities, and generates heat in the same way as exercise. Indeed, shivering can increase heat production up to fivefold in man, but it is both exhausting and a burden on energy reserves. It therefore cannot be sustained over long periods. Moreover, the tremor produced may increase convectional heat loss to the environment, which will reduce the efficiency of the process.

Shivering, together with other components of mammalian thermo-regulation, is ultimately under hypothalamic control. Afferent pathways activating hypothalamic temperature-regulating centres come from thermo-receptors in the skin, hypothalamus, spinal cord, and viscera. These regulate both the onset and extent of shivering, although the relative importance of the different receptors may vary between species. Efferent control is mediated by descending spinal tracts. Despite the fact that all adult homeo-therms can shiver in response to cold, the mechanism is often poorly developed or absent in neonates (see 2.5.2.1). Here, non-shivering thermo-genesis may be of primary importance.

5.2.1.2 *Non-shivering thermogenesis.* Non-shivering thermogenesis is defined as a heat-producing mechanism that liberates energy through processes not involving muscular contraction. It is the process by which the basal metabolic rate is largely maintained. Moreover, it can be increased to raise metabolism two to three times basal levels, and has two main advan-tages over shivering in the newborn. First, shivering could increase air movements around the body surface and so reduce external insulation; this is of particular relevance to the neonate, which has a high surface area to volume ratio. Secondly, shivering could interfere with the fine controlled movements required for the establishment of efficient suckling.

The major site of non-shivering thermogenesis in newborn mammals is brown adipose tissue, a tissue with a marked thermogenic capacity. There are two explanations for thermogenesis by brown fat. One is that mitochondrial phosphorylation is uncoupled from respiration by the presence of large amounts of non-esterified fatty acids. The energy is thus lost directly as heat. The other explanation suggests an increased turnover of ATP via substrate recycling (originally called 'futile cycles'). In brown fat, recycling of triglycerides and fatty acids occurs; lipolysis and reesterification processes result in ATP hydrolysis and heat production. This latter process is stimulated by noradrenaline.

Although it was originally thought that brown fat was present only in neonates, recent studies suggest that it can persist into adulthood. Moreover, cold adaptation (see also 5.2.1.3) may result in the deposition of more brown fat, especially in the dorsal thoracic and lumbar regions. Brown fat also plays an important role in the adult hibernator (see 5.2.3.1).

The regulation of non-shivering thermogenesis in neonatal brown fat is discussed in Section 2.5.2.1. In the adult, other tissues (e.g. liver, skeletal muscle) are also involved in non-shivering thermogenesis, particularly using substrate cycles to generate heat. Here, there is evidence for the involvement of hormones in addition to adrenaline and noradrenaline. Thyroid hormones, for example, are calorigenic, acting to increase ATP turnover by protein synthesis in all tissues. Other hormones with a known calorigenic action include adrenocorticotrophic hormone (ACTH), glucagon, insulin and corticosteroids, but their contribution to normal thermoregulation is uncertain.

5.2.1.3 *Metabolic adaptation and acclimatization.* Although most human inhabitants of cold climates protect themselves with efficient clothing and housing, there is evidence for cold adaptation among some populations. The Alacaluf Indians of Tierra del Fuego, for example, traditionally sleep naked in simple shelters in ambient temperatures of 2–5°C. They do not shiver; instead their survival appears to be due to a basal metabolic rate some 30–40 per cent higher than in other populations. Eskimos exhibit a similar adaptation, although it is not certain whether this is a response to cold or the result of their traditionally high-protein diet (see 5.2.1.4).

Cold acclimation has also been studied. Most non-human homeotherms show a significant increase in non-shivering thermogenesis in response to repeated exposure to low temperatures. Studies on members of polar expeditions have produced evidence for similar metabolic acclimation in man. Figure 5.2 illustrates the effect of acclimation to cold on deep body temperature changes during exposure to standardized cold conditions. The experiments were performed on a group of Australians during a 12-month stay in the Antarctic, and demonstrate an increased metabolic response to cold due to enhanced non-shivering thermogenesis.

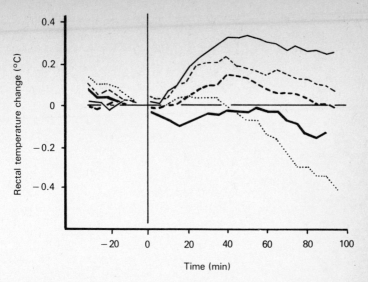

Fig. 5.2 Effect of acclimation to cold on core temperature. A group of Australians were placed in a cold room at time 0 and exposed to standardized cold conditions: (———) in Melbourne, (– – –) after arrival in Antarctica, (———) six months after arrival in Antarctica, (– – –) 12 months in Antarctica, and (------) on return to Melbourne. From Budd, G.M. (1964). *Australian National Antarctic Research Expedition (ANARE) Report Series* B, **IV**, 35. Reproduced in Edholm, O.G. (1978). *Man — Hot and Cold*. Edward Arnold: London.

5.2.1.4 *Increased feeding.* A cold environment stimulates appetite, and an increased food intake in turn elevates metabolic heat production. The increased metabolism is primarily brought about by the protein content of the diet stimulating substrate recycling of amino acids. The high-protein diet of Eskimos, for example, will contribute to their elevated basal metabolism (see 5.2.1.3). Increased metabolism after food consumption is, however, accompanied by an increased blood flow to the peripheral regions of the body. This will increase heat losses (see 5.2.2) and consequently reduce the thermoregulatory contribution of feeding.

5.2.2 Decreased heat loss
5.2.2.1 *Behaviour.* Man's primary behavioural response to cold is one of avoidance. In extreme climates, his schedule of daily activities may reflect his dislike of low temperatures. For example, inhabitants of the Andes in South America rise at dawn and go to bed at sunset. Most of the working day is spent outside, taking advantage of the solar radiation. Andean children are placed in the sun during the day and in communal beds at night. Heat loss in a cold environment may also be decreased by reducing

the surface area available for heat exchange. Hence both animals and man 'huddle' in low temperatures, curling up like a ball with arms folded across the body and legs bent up.

5.2.2.2 *Insulation*. Most terrestrial mammals use fur for insulation. Air is trapped next to the skin and convectional heat loss thus reduced. The insulation value for fur is directly related to its thickness, which is often varied throughout the year to match seasonal temperature changes. Acute decreases in ambient temperature are combated by erection of the hairs (horripilation), which increases the layer of trapped air. Man, the naked ape, has to rely on clothes to provide his insulation (see below). 'Goose pimples' are the most noticeable effect of horripilation in man; low temperatures cause the contraction of the piloerector muscles of the few hairs that he has.

Man's subcutaneous fat will also afford some protection against low temperatures, but there is no evidence for extra insulation in cold-stressed populations. Indeed, Eskimos, Andeans and Alacalufs are all relatively lean. Nevertheless, in general terms, cold-adapted races have a shorter and more compact body shape than their tropical counterparts. Eskimos are traditionally described as showing other 'morphological adaptations' to cold, including eyefold characteristics and facial flatness, but it is unlikely that these will afford any significant protection against cold stress. They also demonstrate a seasonal fluctuation in body weight, with an increase in winter and decrease in spring and summer. However, these changes are too small to affect overall heat balance, and may simply reflect a relative inactivity over the winter months.

Clothing provides man with his main protection against cold. Insulation is provided by air trapped between fabric fibres and between layers of clothing. Thus the traditional clothing of the Arctic is multilayered, with a windproof outer garment. The latter should not be completely impermeable to water vapour, however, or the insulation value of the clothing would be decreased by condensation and possible freezing of evaporated sweat. Insulation also needs to be variable to match man's changing activity. Eskimo clothing, for example, has many vented openings which can be released or closed with drawstrings. Overheating is also prevented by removal of one or more layers of clothing. The extremities of the body present major problems (see 5.2.4.2). The hands, feet and head thus require the most specialized clothing for their protection.

Like clothing, cold-adapted housing must retain heat and be windproof. Man's mastery of fire ensures adequate heat production; heat loss is reduced by the use of compact housing design and suitable insulating materials. The latter have become especially relevant in Western societies recently as a result of rising fuel costs.

5.2.2.3 *Cutaneous vasoconstriction.* Man's primary physiological response to reduced ambient temperature involves a cutaneous vasoconstriction. During acute exposure to cold, vasoconstriction decreases blood flow through the cutaneous circulation and thus reduces convectional heat loss. The vasoconstrictor response to cold stress is due in part to reflexes initiated by both central and peripheral thermoreceptors. These relay information to the hypothalamus, which in turn causes a release of noradrenaline from sympathetic nerves supplying cutaneous blood vessels. Vasoconstriction is also the result of the direct effects of cold on the smooth muscle of the blood vessel wall. This may be due to a cold-induced depolarization of the muscle and/or a delayed removal of noradrenaline at low temperatures.

Although the inital response to cold is that of a cutaneous vasoconstriction, prolonged cooling will result in a paradoxical vasodilatation, with consequent heat loss (the so-called 'hunting' reaction). This response is found primarily in the extremities of the body where little metabolically active tissue exists. Here, vasodilatation alternates with periods of vasoconstriction to prevent tissue damage by severe cold (see Figure 5.3 and Section 5.2.4). Cold vasodilatation represents a general dilatation of all local vessels including an opening of the arteriovenous anastomoses. It is caused primarily by a direct cold-induced paralysis of the peripheral blood vessels, which lose their ability to respond to noradrenaline at low temperatures.

There is some evidence for local acclimation to cold in the fingers and hands of people regularly exposed to low temperatures (fishermen, polar explorers, etc.). Here the initial vasoconstrictor response to cold is less

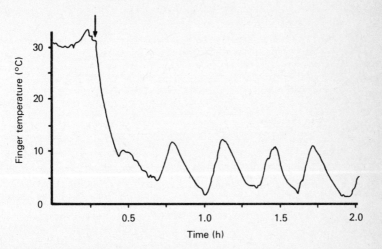

Fig. 5.3   Record of a 'hunting' reaction in a human subject. The subject's hand was immersed in iced water ($-5°C$) at the arrow and finger temperature was recorded over the following 90 min. Room temperature = 15°C; humidity 35%. From Werner, J. (1977). *Pflügers Archives*, **367**, 291–4.

severe, and the onset of the subsequent vasodilatation is more rapid. Eskimos have also been shown to exhibit a higher than normal rate of peripheral blood flow to their extremities when exposed to cold. This is an adaptive response which will prevent cold injury and allow for greater manual dexterity in a cold environment.

A different type of vascular adaptation is seen in the diving women of Korea and Japan (Ama). These women dive for shellfish and edible seaweeds in water temperatures as low as 10°C wearing only minimal clothing. Skin temperatures in their hands and arms appear to be maintained without increasing peripheral blood flow. Instead, a larger proportion of the venous return is channelled through subcutaneous vessels.

5.2.3 *Reduced body temperature*
5.2.3.1 *Hibernation and torpor*. Not all mammals maintain their deep body temperature within the same narrow range as man. Some permit their core temperature to fall to levels approaching ambient temperature, thus reducing the metabolic cost of keeping warm in a cold climate. Heart rate, metabolic rate and other physiological parameters will be correspondingly reduced. Despite this, the animal still retains the ability to rewarm itself to its original temperature without relying on heat from the environment. The phenomenon is known as hibernation, and is a characteristic of small mammals (e.g. hamsters, mice, hedgehogs, squirrels) during cold winter months. Shorter periods of reduced body temperature (e.g. < 24 h) characterize 'torpor'.

Both torpor and hibernation are under precise physiological control. Entry into hibernation is governed by several environmental factors, including ambient temperature, availability of food and water and time of year (light : dark period). It is also thought to have an endogenous (hormonal) component. Arousal is primarily dependent upon ambient temperature, and is a rapid process, frequently raising core temperature through 30°C in 90 min. It is, however, expensive metabolically, employing both shivering and non-shivering thermogenesis (see 5.2.1), the latter involving brown adipose tissue, which is retained in the adult hibernator. Intense sympathetic activity accompanies non-shivering thermogenesis in arousal from hibernation.

5.2.3.2 *Reduced body temperature in man*. The controlled reduction in metabolism and heart rate in human transcendental meditation is similar to that seen during entry into hibernation in other mammals. However, a reduced body temperature in response to a lowered ambient temperature is comparatively rare in man. Nevertheless, there are at least two natural populations which demonstrate the beginnings of a hypothermic response to cold. These are the Australian Aborigines and the Bushmen of the Kalahari Desert in Africa.

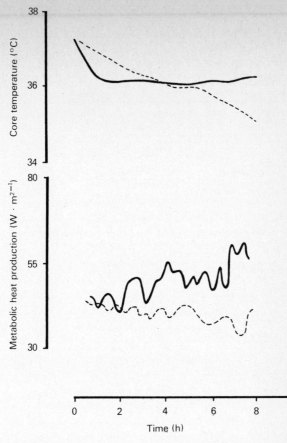

Fig. 5.4 Reduced body temperature in man. Response of a group of male Aborigines (–––) and Europeans (——) to a night of moderate cold exposure. From Richards, S.A. (1973). *Temperature Regulation*, Wykeham Publications, Taylor & Francis: London.

Figure 5.4 shows the response of a group of male Aborigines to a night of moderately cold exposure, and compares their response to that of a group of Europeans. During the night the Europeans maintain their deep body temperature by increasing metabolic heat production. The Aborigines, however, allow their core temperature to fall without attempting to compensate for heat loss by elevating metabolism. A similar response is seen among Kalahari Bushmen. Both ethnic groups are normally subject to high ambient temperatures during the day, and so core temperature would subsequently be raised by solar radiation. The nocturnal hypothermic response may therefore be seen as an adaptation to reduce metabolic costs in natural populations that can be guaranteed a high diurnal ambient temperature.

5.2.4 *Special problems of cold stress*

Special problems of cold stress exist when the normal thermoregulatory mechanisms fail to cope with reduced ambient temperatures. Hypothermia and cold injury are two conditions that fall into this category, and these are considered below. Section 5.2.4.3 examines some of the physiological responses to cold water immersion, which presents thermoregulatory problems in addition to those encountered in air.

5.2.4.1 *Hypothermia.*

The intense shivering that normally accompanies cold stress in man gradually decreases if core temperature drops to 35°C. This is due to an even lower muscle temperature, and signifies the onset of hypothermia. At core temperatures below 35°C, there is further evidence for muscle weakness, with difficulties in walking and general co-ordination. Below 34°C, mental confusion and visual disturbances occur, and consciousness may be lost between 32°C and 30°C. Death may ensue at core temperatures of 28–25°C.

In hypothermia, the most important effects of a reduced temperature are on the heart. Cold slows the pacemaker and cardiac output declines, primarily due to a reduced heart rate. Adequate blood flow to coronary and cerebral circulations may therefore not be maintained. At core temperatures below 28°C, cardiac arrhythmias occur, and ventricular fibrillation is often responsible for hypothermic fatalities at these temperatures. Apart from affecting the heart, hypothermia promotes urinary water loss by suppressing anti-diuretic hormone (ADH) release, and slows respiration by depressing the respiratory centres in the brain. The latter may prove fatal in some mammals, but is thought to be of secondary importance to cardiac disturbances in man.

Recovery from severe hypothermia in air can be rapid and complete. There are, however, additional problems associated with immersion hypothermia (see 5.2.4.3) and in accidental hypothermia in neonates and the elderly.

In the neonate, for example, temperature-regulating mechanisms are not fully developed (see also 2.5). Non-shivering thermogenesis can be used to increase heat production if the ambient temperature falls but the ability to shiver does not develop until the end of the first year in humans. Moreover, the infant is dependent on adult recognition of potential hypothermia as its own behavioural responses are severely limited. This recognition is complicated in humans by the fact that hypothermic babies often have rosy cheeks. Low ambient temperatures shift the haemoglobin dissociation curve to the left and consequently blood in superficial vessels remains undissociated and red. Hypothermia in the elderly is discussed in Section 3.3.3.

Surgically induced local hypothermia is frequently used to man's advantage in operations on organs such as the heart. This reduces the metabolic needs of the organ undergoing surgery, and consequently permits a reduction in its blood supply without the normal adverse effects.

5.2.4.2 *Cold injury*. Frostbite is the most serious form of cold injury, resulting from freezing of the tissues concerned. The severity of the damage produced will depend on both the area and the duration of freezing. In mild cases, only the skin and underlying tissues freeze. If thawing is quickly induced, little damage occurs and the injured skin peels off in a few weeks to be replaced by new growth. However, in severe frostbite, muscle, bone and tendon may also freeze. Damage to the affected cells may be due to the mechanical action of ice crystals, or, more likely, to cell dehydration. Here, ice forming within the cells reduces cellular water content and increases the osmolality. In severe cases, not only is there cellular damage, but also irreversible circulatory changes on thawing. Vasodilatation occurs, and an increased permeability of blood vessels damaged by freezing results in localized oedema. The consequent increase in packed cell volume within the vessels of the affected tissue then reduces or even stops blood flow and gangrene may result with the loss of fingers, toes, or even hands and feet.

A prolonged cooling (rather than freezing) of the lower limbs in cold water or mud results in the condition of 'immersion foot', 'trench foot' or 'peripheral vasoneuropathy'. Such prolonged exposure results in a reduced blood flow due to cold vasoconstriction, and gangrene may sometimes ensue, but the main damage is neuromuscular. Sensory and motor paralysis in early stages of immersion injury are due to direct effects of cold on nerve and muscle. On warming the affected limbs, there is pain, hyperaemia and a varying degree of sensory loss.

Chilblains (erythema pernio) are a milder form of cold injury occurring in fingers, toes or ears. They occur following prolonged cooling of the extremities, especially in individuals with defective circulations. Affected parts become hyperaemic, tender and itchy.

5.2.4.3 *Cold water immersion*. Immersion in cold water presents thermo-regulatory problems in addition to those encountered in air. Indeed, accidental deaths in water are thought to result as much from immersion hypothermia as from drowning. The problems arise because of the higher thermal conductivity of water compared with that of air. Hence the rate of heat loss from the body is greatly increased in an aquatic environment.

Survival time in water varies directly with the water temperature. For example, a naked man will become hypothermic after 20–30 min in water at 5°C and after 1½–2 h in water at 15°C (the mean sea temperatures around the coast of England and Wales during January and July, respectively). Increased insulation, of course, will prolong survival times. However, neither fur, as used by animals, nor clothes, as used by man, are as effective as insulators in water as they are in air. Both normally insulate by trapping pockets of air, and these will be lost, or at best compressed, in water.

Subcutaneous fat provides much more effective insulation in an aquatic environment. Indeed, many aquatic mammals (e.g. seals and whales) use blubber effectively to provide insulation in cold waters. Terrestrial mammals including man have smaller subcutaneous fat deposits in order to facilitate mobility on land. Nevertheless, man's fat layers will decrease heat loss in water, and thin men become hypothermic in water more rapidly than fat men.

Increased movement in cold water will increase heat loss still further by disturbing the layers of warmer water in immediate contact with the body surface. Hence those in potential danger of shipwreck in cold waters are advised to float with a life-jacket or wreckage and await rescue rather than attempt to swim any great distance to shore.

5.3 Responses to high temperatures

Man appears well adapted for life in a tropical environment because of his ability to sweat (see 5.3.2.1). However, very few natural populations live in climates where air temperature exceeds deep body temperature by more than 12°C. This contrasts strongly with man's ability to survive in very cold climates, where ambient temperature may be more than 80°C below core temperature (see 5.2). Thus, in man, overheating appears to be more critical than overcooling. Indeed, death may ensue with a rise in core temperature of only 5°C as against a fall of 10°C.

Man's responses to heat are both behavioural and physiological. High temperatures are avoided, and suitable clothing and housing adopted. Food intake is reduced. Physiological responses include sweating and cutaneous vasodilatation. Acclimatization to heat is much more evident than similar responses to cold.

Heat responses fall into three main categories: decreased heat gain (both from endogenous production and external sources), increased heat loss, and a limited tolerance of hyperthermia. The latter is characteristic of mammals other than man and is therefore discussed only briefly.

5.3.1 *Decreased heat gain*
5.3.1.1 *Reduced activity*. A general apathy and inertia are synonymous with hot climates. Both serve to decrease endogenous heat production. If manual work must occur, it is usually confined to the cooler parts of the day. Some desert mammals reduce their activity still further in summer months, remaining dormant in their burrows. This type of dormancy, in response to high ambient temperatures, is known as 'aestivation' in contrast

to 'hibernation' (see 5.2.3.1), although the distinction between the two processes is fine.

5.3.1.2 *Reduced feeding*

Just as a cold environment stimulates appetite, so a hot environment promotes anorexia. The decreased food intake in turn reduces endogenous heat production by its effect on metabolism. Food preference is directed towards items with a high water content (e.g. fruit and salad vegetables).

5.3.1.3 *Reduced heat gain from the environment*

Behavioural adaptations are extremely important in reducing heat gain in a hot climate. Hence most mammals (mad dogs and Englishmen excepted!) avoid the intense heat of the midday sun. Tropical man typically starts work early, remains sedentary around noon, and engages in less strenuous activities in the shade during the afternoon.

Suitable clothing and housing provide man with insulation against solar radiation. However, it is difficult for man to keep out external heat using insulation without also preventing loss of his own metabolic heat. This may in part explain why overheating is a more critical problem than overcooling in man and other species (see 5.3).

The use of clothing varies between areas of dry and humid heat. In humid tropical regions, minimal use of clothing reduces the body's heat load and increases the surface area for sweat evaporation, which is normally impeded in such conditions (see 5.3.2.1). In dry desert areas, however, clothing plays an important role both in reducing heat gain from the environment and facilitating heat loss from the body. This is achieved by the use of loose-fitting, lightweight, light-coloured materials. These reflect much of the short-wave solar radiation and also allow for the circulation of air necessary for effective evaporation of sweat. Clothing will also prevent skin damage by burning (see 5.3.4.4) and protect against hot desert winds.

Housing in hot dry areas is constructed with the same objectives in mind as desert clothing. Heat gain from the environment is minimised by the use of compact-design houses with light-coloured exteriors, often semi-subterranean in construction. Heat generating sources (cooking fires, etc.) are kept away from the main dwelling, and particular attention is given to controlled ventilation. One real problem in hot dry climates is the large diurnal temperature variation resulting from the lack of atmospheric moisture and cloud cover. Diurnal ranges of 35°C are commonly recorded in continental hot deserts. Azizia in Tripoli, North Africa, once recorded 52°C and −3°C during 24 h, a record diurnal range. In such conditions, building materials with a high heat capacity (e.g. stone, mud, etc.) are useful because they can absorb heat in the day and radiate it at night.

5.3.2 *Increased heat loss*

5.3.2.1 *Sweating*. Sweating is man's major means of effecting an increased heat loss in a hot environment. Indeed, man can produce more sweat per unit area of skin than any other mammal. An adult man working in hot dry conditions, for example, may secrete 12 litres of sweat per day. During shorter periods of muscular exertion, this rate may approach 3 litres $\cdot$ h^{-1}. Evaporation of sweat is necessary for heat to be lost. The latent heat of evaporation of water at 37°C is 2.4 kJ $\cdot$ ml^{-1}. If evaporation occurs on the skin, the majority of this heat will come from the body itself. For a maximum sweat rate of 3 litre $\cdot$ h^{-1}, heat loss will therefore approximate 7000 kJ $\cdot$ h^{-1}. Assuming a basal metabolic rate of 350 kJ $\cdot$ h^{-1}, sweating allows man to lose heat up to twenty times as fast as it is produced by basal metabolism. The rate of evaporation of sweat, however, is dependent upon the amount of water vapour in the atmosphere. At 100 per cent humidity, for example, no evaporation will occur, and so this method of losing heat will not be available. Atmospheric humidity is thus an essential consideration in any discussion of the role of sweating.

Sweat itself is a hypotonic solution produced by eccrine glands in the dermis (see Figure 5.5). The deep subdermal coiled portion of the gland is responsible for the production of a primary secretion. This is subsequently modified by solute reabsorption as the fluid moves along the duct towards the skin surface. The main components of sweat are water and sodium chloride, together with smaller amounts of potassium, calcium, urea, uric acid and ammonia. At low rates of secretion, the sodium chloride content of sweat reaching the skin surface is low (5 mmol $\cdot$ litre^{-1}), whereas at high secretion rates there is less time for reabsorption and the sodium content can approach 70 mmol $\cdot$ litre^{-1}.

Like other thermoregulatory responses, sweat production is regulated by the hypothalamus in response to input from cutaneous and central thermoreceptors. Secretion of sweat on to the skin surface is under sympathetic control; cholinergic sympathetic nerve fibres end on or near the glandular cells.

Although man makes effective use of sweating as a thermoregulatory process, many smaller mammals cannot. For mammals to thermoregulate in hot climates, water must be evaporated in proportion to the heat load. The latter is far greater in small animals, since their surface area to volume ratio is higher, and this would therefore have to be matched by very high sweat rates and excessive fluid losses. Body size and shape are also important in man. The 'ideal' body type for desert conditions is tall with long, lean limbs and low subcutaneous fat. Tallness maximises the surface area to body weight ratio for evaporative cooling; leanness facilitates heat conduction from deep body tissues. The Nilotic peoples of the Sudan are often cited as an example of optimal human form for desert environments.

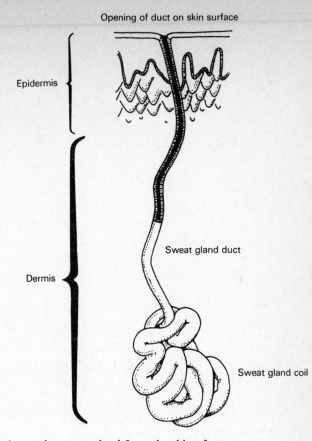

Opening of duct on skin surface

Epidermis

Dermis

Sweat gland duct

Sweat gland coil

Fig. 5.5 An eccrine sweat gland from the skin of man.

A critical factor in the use of sweating to combat a high heat load is the availability of water. Where drinking water is unrestricted, a balance is soon established between fluid intake and the combined fluid losses of sweating and urinary excretion. Where water is not immediately available, varying degrees of dehydration ensue. Normally, in man, thirst mechanisms are triggered with a fluid loss of 2 per cent of the body weight. Urinary water loss is also reduced by increased ADH secretion. The lethal limit for human dehydration is a fluid loss of 15–25 per cent of the body weight, a figure similar to that seen for a number of other mammals.

Sodium chloride is the major solute in sweat. Excessive sweating will therefore also result in salt loss. If a negative salt balance develops, increased levels of aldosterone act on the renal distal tubules and sweat gland ducts to conserve sodium. Section 5.3.4.3 considers in more detail the specific problems of salt-deficiency and water-deficiency heat exhaustion.

5.3.2.2 *Sweating and acclimatization.* Heat acclimatization in man is much more evident than similar responses to cold. Indeed, all human populations acclimatize to desert conditions within one or two weeks. An increased sweat rate is central to the acclimatization, and is accompanied by a diminished heart rate and lowered core temperature. The latter two are both raised on first exposure to heat (see 5.3.2.4). Acclimatized individuals also show a decreased salt concentration in sweat, which prevents some of the symptoms of salt-deficiency heat exhaustion (e.g. muscular cramps — see 5.3.4.3), and a reduced urine production (see also 5.3.2.1). Figure 5.6 illustrates the phenomenon of heat acclimatization and shows the changes in sweat rate of a group of volunteers exposed in a hot room for four periods of 5 or 6 consecutive days. After 20 days of exposure, the sweat rate is double that on day one.

Humans adapted to heat thus sweat more, and they also start sweating at a lower threshold core temperature. In natural environments the increase in sweating is much more marked in acclimatization to hot humid conditions than to hot dry ones. Sex and ethnic differences are also apparent. In all natural populations studied, for example, men have a much higher sweat rate than women (often twice the value for comparable groups). This in part

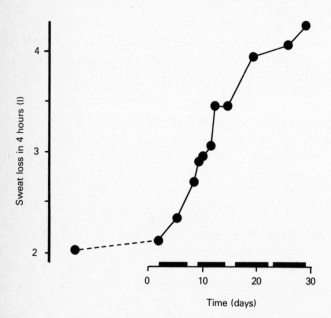

Fig. 5.6   Sweating and acclimatization. Increased sweat rate of a group of subjects exposed in a hot room for four periods of 5 or 6 consecutive days. Exposure days shown by horizontal black bars. From Edholm, O.G. (1978). *Man — Hot and Cold*. Edward Arnold: London.

reflects the fact that men are usually more physically active than women, but endocrine factors may also be involved. Controlled hyperthermic experiments also reveal striking differences between indigenous populations of hot countries. Native tribes of New Guinea, for example, have much lower sweat rates than either Nigerians or Europeans when studied under similar conditions.

5.3.2.3 *Panting and salivation.* In man, evaporative heat loss is almost exclusively the result of sweating, but this is not true for all mammals. The dog, for example, has few sweat glands and cools primarily by panting, which increases evaporation from the upper respiratory tract. Respiratory frequency increases tenfold in a panting dog; man, however, shows only a modest increase in ventilation if heat stressed.

There are several advantages of panting over sweating as a means of losing heat. First, the panting animal provides its own air circulation over moist surfaces, thereby facilitating evaporation. Sweating, of course, is always reliant on external conditions. Secondly, panting does not result in the inevitable loss of sodium which is characteristic of sweating. Thirdly, venous drainage from the nasal passages of some panting animals serves to pre-cool arterial blood flowing to the brain. Such animals can tolerate a higher body temperature and still maintain a normal brain temperature (see 5.3.3). Disadvantages of panting result primarily from the increased ventilation which will increase the work of breathing and causes alkalosis.

A few mammals spread saliva over their fur as an additional evaporative heat loss mechanism. This approach, however, is neither widespread nor very effective.

5.3.2.4 *Cardiovascular responses.* Increased ambient temperature produces a rapid cutaneous vasodilatation which promotes the transfer of heat from deep to superficial body tissues and thence to the environment. Both central and peripheral thermoreceptors are involved in the response and the enhanced skin blood flow is achieved by the opening of arteriovenous anastomoses and superficial veins. At the same time, blood flow to regions such as the kidney and gut are reduced.

As core temperature increases, cardiac output also rises, but blood pressure changes are minimised by the fall in total peripheral resistance produced by the vasodilatation described above. The raised cardiac output is due initially to an increased heart rate, with temperature having a direct effect on the sino-atrial node. If heat stress is prolonged, however, an increased stroke volume may also occur.

During acclimatization, some of the above changes are reversed. Heart rate falls as core temperature is lowered, and circulatory adjustments are balanced by an increase in blood volume.

5.3.3 *Hyperthermia and heat storage*

Several mammals other than man have an additional heat response inasmuch as they can tolerate a certain degree of hyperthermia. Such animals allow their core temperature to rise in hot environments and thereby economize on evaporative cooling. The camel, for example, retains heat and makes no attempt to thermoregulate until its core temperature exceeds 40°C. This may result in the saving of up to 5 litres of sweat in a day, which will be particularly important if water supplies are restricted.

Some exercising mammals use similar methods; heat is retained during exercise and dissipated at rest. Fast-running African gazelles, for example, can tolerate a core temperature of 46°C for several hours if water is limited. The brain temperature of these animals is kept much lower (around 40°C) by a special heat-exchange network at the base of the brain. Here, the carotid artery divides into a network of small arteries (carotid rete), which lie in a large venous sinus containing cooled blood returning from the nasal passages. Arterial blood is thereby pre-cooled before it enters the brain.

5.3.4 *Special problems of heat stress*

Special problems of heat stress occur when the body fails to cope with increasing environmental temperatures. These range from relatively minor heat-induced skin disorders (5.3.4.1) to more serious general conditions (5.3.4.2–5.3.4.4).

5.3.4.1 *Heat-induced skin disorders*.

Sunburn and prickly heat are two heat-induced disorders affecting the skin itself. Sunburn results from the direct exposure of the skin to solar radiation. In mild forms, it is characterized by an increased production of melanin, which gives the skin a 'tan' and acts as a natural protection from the sun's rays. In severe forms, skin damage is caused, with erythema, pain, blistering, and subsequent peeling off of the epithelium.

Prickly heat (miliara rubra) is characterized by the appearance of minute vesicles caused by a blocking or narrowing of the sweat gland ducts. Sweat is therefore unable to reach the skin surface and a localized swelling results. The vesicles are extremely itchy, and scratching can cause further epithelial damage.

5.3.4.2 *Heat syncope*.

Heat syncope or collapse is characterized by fatigue, dizziness and a temporary loss of consciousness under hot conditions. During this time, blood pressure is low and the pulse usually weak. Such symptoms are, however, transient and recovery is usually rapid when the subject is removed to a cool environment or simply lies down. The condition is caused by a vascular collapse due to pooling of blood in skeletal muscles and skin, particularly in the lower limbs. It is prevented by proper acclimatization, since the increased blood volume which occurs on repeated exposure to heat counteracts the tendency for peripheral pooling.

5.3.4.3 *Heat exhaustion.* Mild cases of heat exhaustion are similar to heat collapse. More severe cases have varying symptoms depending upon the cause of the condition. Water-deficiency heat exhaustion, for example, is caused by insufficient water replacement of fluid losses. The characteristic order of the symptoms of dehydration in a hot environment is shown in Figure 5.7. Most people can tolerate a 3–4 per cent water deficit. With a loss of 5–8 per cent, however, fatigue sets in; and a loss of over 10 per cent causes physical and mental deterioration. Water deficiency primarily affects the extracellular fluid volume but, if continued, fluid is also withdrawn from intracellular compartments. A gradual decrease in plasma volume accompanies the dehydration, and this is responsible for the initial symptoms. Later, cellular damage results from the accompanying increase in plasma osmolality.

Salt-deficiency heat exhaustion occurs when sodium chloride losses in sweat are not adequately replenished by salt in the diet. Tissue osmolality consequently falls, and extracellular compartments contract. The syndrome is characterized by heat cramps, usually in the legs, arms or back, and by fatigue and dizziness.

5.3.4.4 *Heat stroke.* Heat stroke is a much more serious condition, resulting from a complete loss of thermoregulatory control in the heat. Initial characteristics are a loss of energy and irritability, but there then follow serious neurological and mental disturbances. Sweating diminishes or ceases and the subject passes into a coma as core temperatures rise to 42°C or beyond. Unless treatment is started immediately to lower the body temperature, death usually ensues.

The loss of thermoregulatory control in heat stroke appears to be due primarily to a failure of the sweating mechanism. This may be a complete failure resulting from hypothalamic malfunction, or simply an inadequacy to meet the needs of the particular heat load. Cellular degeneration and coagulation of proteins both accompany high core temperatures, and irreversible changes are effected.

5.4 Further reading

Adolph. E.F. (1947). *Physiology of Man in the Desert*. Interscience: New York.

Edholm, O.G. & Weiner, J.S. (1981). Thermal physiology. In: *The Principles and Practice of Human Physiology*, eds O.G. Edholm & J.S. Weiner, Chapter 3. Academic Press: London, New York.

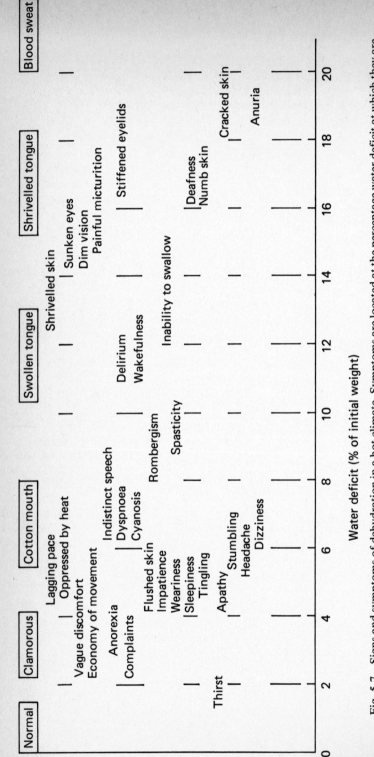

Fig. 5.7 Signs and symptoms of dehydration in a hot climate. Symptoms are located at the percentage water deficit at which they are normally first seen. From Adolph, E.F. (1947). *Physiology of Man in the Desert*. Interscience: New York.

Hensel, H. (1981). *Thermoreception and Temperature Regulation*. Monographs of the Physiological Society No. 38. Academic Press: London, New York.

Keatinge, W.R. (1969). *Survival in Cold Water*. Blackwell: Oxford.

Moran, E.F. (1979). *Human Adaptability: an Introduction to Ecological Anthropology*. Duxbury Press: Massachusetts.

Robertshaw, D. (ed.) (1979). *Environmental physiology III. International Review of Physiology* Vol. 20, Chapters 1–4. University Park Press: Baltimore.

Rolls, B.J. & Rolls, E.T. (1982). *Thirst*. Cambridge University Press: Cambridge.

Chapter 6

Altered time

Summary

Humans are rhythmic animals. Possession of circadian ('about a day') rhythmicity enables them to fit better into a rhythmic environment. This rhythmicity derives from two causes, one of which is the direct effects of the environment and our habits whereby we eat, drink and are more active when awake but rest and fast when we are asleep. The second cause is an internal 'clock' which, at least in rodents, is probably found in the hypothalamus. This internal clock is adjusted to an exact period of 24 h by the rhythmic environment and in man the factors responsible for this are mainly social ones, for example: knowing when it is appropriate to eat, to shop, to make noise, to arrange meetings, etc.

A knowledge of circadian rhythmicity is relevant to medicine mainly in diagnosis and chemotherapy. First, it implies that the range of a given parameter might depend upon the time of day when it is measured and this might influence when diagnosis takes place. Secondly, it now seems that the efficacy of a drug (the ratio of its therapeutic to toxic effects) might show circadian rhythmicity. This challenges the physician to attempt to optimise his use of drugs, not only with respect to the dose, but also with respect to the time(s) of administration of the drug.

The internal clock is slow to adjust to changes in schedule and so, for some days after time-zone transition and during night work, there is a 'mismatching' between body time and environmental time. This is associated with a general malaise, sometimes known as 'jet-lag syndrome', which eventually causes some night workers to leave shift work. In the case of time-zone transition, this malaise is transient, since the internal clock adjusts after about a week to the new environment. However, in the case of the shift worker, adjustment is slower and often incomplete; this reflects the conflicting demands of his work, his family and social influences in general.

6.1 Introduction

6.1.1 *Homeostasis and rhythmicity*
When a physiological parameter such as deep body temperature is studied over the course of 24 h, it is observed not to be constant throughout. This is shown, together with other examples, in Figure 6.1. These parameters all show higher values during the daytime than during the night, although the time of peak values can vary between individuals and between parameters. These differences are not caused by errors in the homeostatic feedback systems but rather indicate changes in their 'set point' during the course of the 24 h. Thus deep body temperature is *controlled* at a higher value during the daytime than during the night.

6.1.2 *Endogenous and exogenous components*
It might be argued that such rhythmicity simply reflects external rhythmic factors which affect the parameter under consideration. For example, daytime activity and feeding might be expected to raise blood pressure, temperature and urinary potassium excretion whereas sleep (associated with inactivity and fasting) would lower these parameters. It could further be argued that the higher nocturnal plasma concentrations of some hormones (e.g. growth hormone, prolactin, cortisol) reflect greater growth and repair taking place at this time of inactivity. This argument can be tested by submitting volunteer subjects to a 'constant routine'. The subjects are kept awake and sedentary for 24 h in an environment from which all rhythmic cues have been removed (i.e. an environment of constant lighting, temperature, humidity and noise). They take an identical snack each hour,

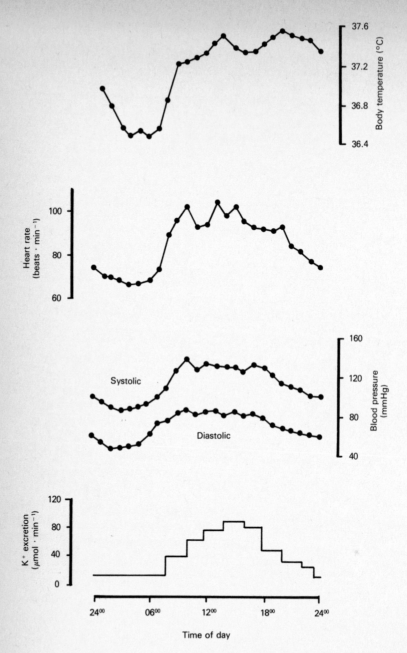

Fig. 6.1 Some examples of circadian rhythms from subjects who slept at night (midnight–08.00) and ate, drank and were active during the day.

which maintains their normal intake of the major electrolytes and provides some nourishment.

In Figure 6.2 is shown the effect of this constant routine on rectal temperature in a group of eight male subjects, together with the results in the same subjects during a normal 24-h period, that is, when they slept at night. The following points can be inferred from this figure.

(1) The rhythm does not disappear during the constant routine. This is attributed to some internal factor, or factors, called the internal 'clock' or endogenous component.

(2) The rhythm during conventional hours of sleep and wakefulness differs slightly from that during the constant routine, showing higher maximum and lower minimum values. This difference is attributed to the environment and daily routine and is called the exogenous component.

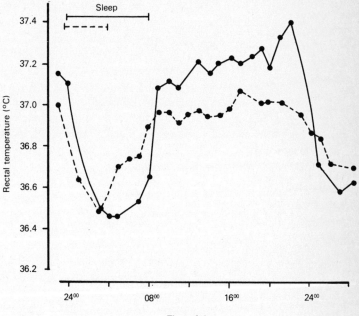

Fig. 6.2 Mean circadian changes in rectal temperature measured hourly in eight subjects living a normal nychthemeral existence (——) and in the same subjects during a constant routine in which they were woken at 04.00 and spent the subsequent 24 h awake in constant light and taking hourly small identical snacks (–––). From Minors, D.S. & Waterhouse, J.M. (1981), *Circadian Rhythms and the Human*. John Wright: Bristol.

(3) The endogenous and exogenous components are in phase. That is, during the daytime, the internal clock, the external environment and the daily routine combine to increase temperature: at night they all tend to lower it.

6.2 Physiological concepts

This division of a rhythm into endogenous and exogenous components applies in principle to all parameters, even though the causes of the exogenous component might differ. Thus, for the concentrations of amino acids in plasma, the exogenous component might reflect the rhythm in feeding; for plasma growth hormone concentration, that in sleeping; and for rectal temperature, that in activity.

6.2.1 *The endogenous component*
A more detailed study of the internal clock requires data to be collected for more than 24 h. Longer experiments involving a constant routine are possible but the effects of fatigue then become intrusive. A more satisfactory method is to allow subjects to retire, rise and eat as they wish but in the absence of any external cues as to the real time. Subjects record their actions in such a way that the real time at which these occurred can be recorded by the experimenters. Such experiments have been performed using three types of time-free environments: caves, the circumpolar regions during summer, and specially-constructed 'isolation units'. Caves and the polar regions are both uninviting environments and, because of their inaccessibility, limit the kind of experimentation that can be carried out. By contrast, an isolation unit consists of an experimental chamber offering conventional accommodation and facilities but is constructed so that noise, rumble from traffic and other cues (e.g. changes in the power supply from the national grid) do not reach inside. Such a unit can much more easily be incorporated into a conventional laboratory and enable more data to be obtained.

The experimental findings from all three environments concur. In all cases, subjects continue to show rhythmicity which remains in phase with their pattern of sleep and wakefulness. (Some exceptions will be mentioned later.) However, the periodicity of the rhythm is not exactly 24 h: it is closer to 25 h and so, after the passage of about 72 real days, only approximately 69 'days' will have been experienced by the subject.

6.2.2 *Entrainment to an exact 24 hours*
Time-free environments establish the presence of an internal circadian clock (*circa*, about: *diem*, day). However, they also raise a problem. If the clock is to be of use in an environment dominated by the sun, some form of

adjustment to a 24-h period is necessary. This synchronization or entrainment is achieved by rhythmic factors in the environment termed synchronizers or (from the German) zeitgebers.

The nature of the zeitgeber varies between species. For plants, it is the alternation of night and day; for shore-dwelling creatures (entrained to a lunar day), the rhythmic buffeting from high tides; and for many terrestrial animals, the availability or otherwise of food. For man, there is still some controversy. Claims for rhythms in feeding, social interaction and light have all been made. Probably all these factors exert some effect. Thus we go to bed even if we do not feel particularly tired because we know we have to get up the next morning, and much of our lifestyle (mealtimes, shopping excursions, when not to make a noise, etc.) is timed to fit in with the lifestyle of others. It is this mass of rhythmic information from our environment that acts to synchronize our natural 25-h circadian rhythm to a 24-h period. It is important to realize that, under normal circumstances, all these environmental influences act in concert to constrain us to be awake and active during the daytime and to sleep at night.

In summary, therefore, body rhythms consist of two components: an endogenous component produced by an internal clock, and an exogenous component produced by a rhythmic lifestyle and environment. Unadjusted, the internal clock runs slow, with a circadian period of about 25-h but normally it is entrained (synchronized) to a 24-h day by environmental zeitgebers (synchronizers). In man, the main synchronizer is the rhythmic changes in social influences.

6.2.3 *Relative size of endogenous and exogenous components*

It can be concluded that the environment exerts important influences upon circadian rhythms both by contributing to the exogenous component of a rhythm and by entraining the endogenous component. In the case of rhythms with large exogenous components, this effect of the environment can become dominant and give misleading information about the endogenous component. This important point can be illustrated by a study in which subjects lived a 28-h day, that is, their routine was governed by watches that showed the passage of a full 'day' only when 28 real hours had passed. Figure 6.3 shows that, in one such subject, heart rate showed a 28-h rhythm which was due to the marked exogenous component present in this rhythm. By contrast, oral temperature showed a period much closer to 24 h, due to its following the internal clock. Thus, over the course of 7 (real) days of the experiment, temperature showed seven full cycles and heart rate only six. As a result, the two rhythms were completely out of phase with each other on days 3–4. By assessing the relative importance of the exogenous (28 h) and engodenous (near-24 h) components in this kind of experiment, it is possible to rank variables in terms of the relative importance of

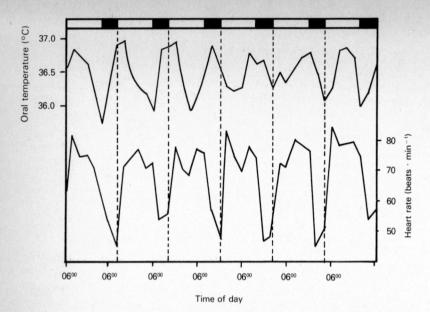

Fig. 6.3   Oral temperature and heart rate of a subject living on a 28-h routine. Data represent 4-hourly means from three 7-day cycles of the 28-h routine. The horizontal axis is divided into real days and the dashed vertical lines represent the ends of 28-h periods. Black bars represent the times the subject was asleep. Note that a peak in oral temperature occurs for every real day whereas for heart rate a peak occurs on each artificial day. From Kleitman, N. & Kleitman, E. (1953). *Journal of Applied Physiology*, **6**, 283–91.

endogenous and exogenous components. It was in this way that the data of Table 6.1 were obtained.

6.2.4 *The site of the clock*

The complex interactions between physiological parameters in response to a variation in the environment are the common theme of chapters in this book. It can be extended to their circadian rhythms. Figure 6.4 shows some of these interactions. Such a scheme has two important implications. First, some rhythms that are comparatively difficult to measure (e.g. mental performance) might be inferred from others that are easier to assess (e.g. temperature, plasma and urinary catecholamine concentrations). Secondly, when this scheme is compared with the data of Table 6.1, it can be seen that those rhythms with high endogenous components tend to be associated with hypothalamic activity. This result is believed to offer some clues as to the whereabouts of the internal clock, which cannot be investigated directly in man. In rodents and primates the results of the surgical removal of tissues indicate that the suprachiasmatic nuclei of the hypothalamus may be

Table 6.1. Common circadian rhythms grouped according to the approximate relative importance of endogenous and exogenous components. Group A, highest endogenous component through to Group D, lowest.

Group A
   Plasma adrenocorticotrophic hormone concentration
   Plasma cortisol concentration
   Rapid eye movement sleep
   Deep body temperature

Group B
   Urinary adrenaline excretion
   Urinary potassium excretion

Group C
   Urinary sodium excretion
   Mental performance
   Airway resistance

Group D
   Heart rate
   Plasma amino acid concentration
   Plasma growth hormone concentration
   Stage 4 sleep

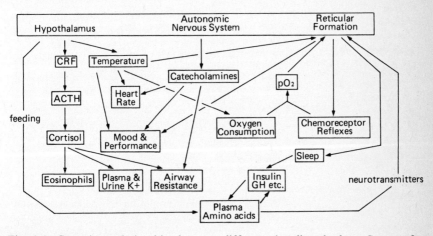

Fig. 6.4   Some interrelationships between different circadian rhythms. See text for further details. From Minors, D.S. & Waterhouse, J.M. (1981). *Circadian Rhythms and the Human*. John Wright: Bristol.

involved. Further, recordings of electrical activity from tissue slices containing these nuclei (but not from other areas of the brain) have shown circadian rhythms in constant conditions. However, other experiments performed *in vivo* have indicated that not all rhythms disappear after removal of the suprachiasmatic nuclei. Therefore, the precise role to be ascribed to the suprachiasmatic nuclei (which also exist in man) has yet to be established. These last experiments raise the possibility that more than one internal clock exists and this will be discussed further later (see 6.2.5).

Another recent finding is the presence in mammals of a neural pathway which directly links the retina and the suprachiasmatic nuclei. It is believed that entrainment of the internal clock to a light–dark zeitgeber takes place through this pathway. Schemes can also be devised for suggesting how feeding cycles can act as zeitgebers (via increases in plasma amino acid concentrations and brain neurotransmitter synthesis) but their relevance to man remains speculative.

6.2.5 *Is there only one clock?*

The use of experiments in time-free environments has already been described. In about one-third of such experiments it is observed, especially with more prolonged isolation, that 'spontaneous internal desynchronization' takes place (Figure 6.5). In this figure, normal rhythms of sleep and activity and of rectal temperature are seen for the first 14 days (see 6.2.1). The rhythms have a period of about 26 h and are in phase with each other, peak temperatures being associated with activity and minimum values with sleep. However, from day 15 onwards, even though the temperature rhythm continues very much as before, that of the sleep and activity shows a new period of about 33 h. As a result of this, peak temperatures no longer reliably coincide with activity, nor minimum values with sleep. This result strongly suggests that two clocks are present, an inference already drawn from studies involving surgical removal of the suprachiasmatic nuclei. If a second clock is present, then its site, properties and the zeitgebers that entrain it are quite unknown. It is presumed that the two clocks are normally linked together but what might happen during changes in the sleep–wakefulness routine as a result of time-zone transition (see 6.3.4) and shift work (see 6.3.5) is unknown.

6.2.6 *Development during infancy*

Circadian rhythmicity is absent from the neonate and develops gradually during the course of the first 5 years. This might suggest that the acquisition of a clock is the result of a rhythmic environment but, if so, why a clock with a period nearly always more than 24 h? The importance of the environment is supported by the observation that circadian rhythms are poorly developed in blind persons and in Eskimos. However, this might

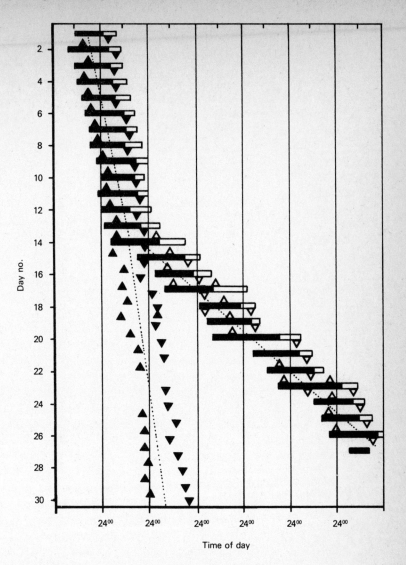

Fig. 6.5 Circadian rhythms of wakefulness and sleep (black and white bars respectively) and of rectal temperature (▲ maxima, ▼ minima) in a subject who lived alone in an isolation unit without indication of the time of day. Successive days are plotted from above down. From days 1 through to 14 the subject is internally synchronized with the two rhythms showing similar periods ($t = 25.7$ h). At day 15, spontaneous internal desynchronization takes place so that thereafter the two rhythms show different periods (rectal temperature $t = 25.1$ h; sleep-wakefulness $t = 33.4$ h). Open triangles show positions of temperature maxima △ and minima ▽ in relation to the rhythm of wakefulness and sleep. From Wever, R. (1975). *International Journal of Chronobiology* **3**, 19–55.

reflect a weak exogenous component to the rhythm, in turn resulting from irregular sleep–waking rhythms. Thus, the circadian rhythms in Eskimos are more marked at times of the year when the community adheres more strongly to a structured routine. By contrast, the view that the clock has a genetic component comes mainly from breeding experiments with *Drosophila* (fruit fly) species: mutants that show an abnormal rhythmicity transmit it to their progeny according to Mendelian laws of inheritance. In humans, the evidence for a genetic component is weaker but ethical considerations severely limit experimentation. Rhythmicity develops in infants fed on demand and there is even a report that rhythmicity developed in an infant reared in an environment of constant light and fed on demand. In these cases it can be argued that the developing rhythmicity has not been imposed from the environment. However, two other factors must be considered. First, it is unlikely that the external environment was absolutely constant; rhythms in noise, routine and availability of attention are difficult to remove completely. Secondly, the child has lived before birth in an environment with rhythmic changes of temperature, pressure (maternal posture) and hormone concentration. The fetus is known to show circadian rhythms in movement during at least the last 3 months of pregnancy and so it might be *before* birth that a rhythmic environment exerts its effect upon the individual.

Perhaps the development of the circadian pacemaker system is like that of the visual cortex as demonstrated in kittens. A genetic component is present that initially controls the properties of cells in the primary visual cortex; these cells are then influenced by the visual environment during the first few months of life. By the time the kitten has grown up, these properties of the visual system have become relatively stable.

In summary, the relative importance of the endogenous components depends upon the parameter under consideration. Those parameters with high endogenous components, which are therefore considerably influenced by the internal clock, are believed to give some clues as to where that clock might be. This indirect reasoning leads to a consideration of the hypothalamus, and experiments upon other mammals suggest an important role for the hypothalamic suprachiasmatic nuclei. However, the position is complicated by evidence that indicates the presence of more than one clock. There seems reason to believe that the normal development of circadian rhythmicity involves genetic and environmental influences.

6.3 Implications and applications

6.3.1 *Usefulness*
The general usefulness of an internal clock in plants and animals is not hard to understand. For example, plants could position their leaves in readiness

for sunrise; nocturnal animals would know when to leave the safety of their burrows. In man these considerations are not appropriate but the clock still has a usefulness: it prepares an individual both for the rigours of a new day and, later on, for the next sleep. Thus, from about 04.00 onwards, body temperature, blood pressure and the plasma concentrations of adrenaline and cortisol rise to prepare the individual for activity when he wakes. Similarly, as the evening progresses, plasma adrenaline concentration, mental processes and body temperature all begin to decline in preparation for sleep (Figure 6.1). In other words, the internal clock prepares the individual for future needs. This preparation, together with the adjustment brought about by zeitgebers, enables the rhythmic organism to fit stably into a rhythmic environment.

6.3.2 *Medical applications*
Having looked at some of the physiological concepts underlying circadian rhythmicity, we shall now look at the practical consequences. We shall begin in the clinical field by asking whether a knowledge of circadian rhythmicity aids in the diagnosis and treatment of disease. This introduces a very new field and the progress made so far is modest.

6.3.2.1 *Diagnosis and prediction of illness*. Diagnosis of illness is often

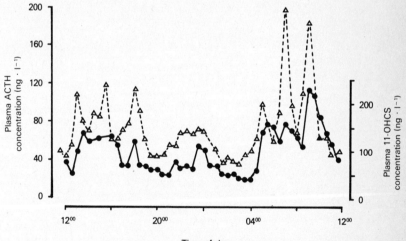

Fig. 6.6 Circadian variations of plasma cortisol (11-OHCS, ●——●) and adrenocorticotrophic hormone (ACTH, △---△) concentrations in a female subject, determined by half-hourly sampling over a 24-h period. From Krieger, D.T., Allen, W., Rizzo, F. & Krieger, H.P. (1971). *Journal of Clinical and Endocrinological Metabolisum* **32**, 266–84. © 1971 The Endocrine Society.

assisted by showing that some parameter has a value outside the range compatible with health. The existence of a range is due partly to variations between normal individuals; this can often be reduced by taking into account factors such as sex, age, weight, etc. Such considerations are sufficient for most parameters but are inadequate when one with a circadian rhythm of considerable amplitude is considered. An example is disordered secretion from the adrenal cortex. The normal rhythms of cortisol and adrenocorticotrophic hormone (ACTH) secretion are shown in Figure 6.6. This figure indicates that a case of suspected adrenal insufficiency would be better diagnosed from blood samples taken after waking (when values should be higher) than from samples taken in the evening. Similarly, suspected Cushing's disease (overproduction of cortisol) is better assessed in the evening.

Examples as clear as this are uncommon, but in each of the following, the timing of diagnosis can be important.

(1) Skin (especially forehead) temperature better assesses fever in the morning than in the evening. This is because body temperature is falling in the evening (Figures 6.1 and 6.2) and so heat loss through the skin will normally be highest then. In the morning the body is conserving heat and so skin temperature should be lower.

(2) The release of growth hormone is normally associated with the early part of sleep and, except during puberty, is usually very low during the daytime. Accordingly, assessment of secretory excess is better made in the daytime and of deficit during the first half of a night's sleep.

(3) Ventilation normally falls during sleep and so pulmonary CO_2 concentrations rise and O_2 concentrations fall slightly. In the case of some microfilarial parasites, these changes seem to cause the parasite to leave the lungs and circulate in the blood (so allowing it to be detected in a blood sample). However, the problem is exacerbated by the observation that different species leave the lungs and travel round the blood at different gas tensions.

(4) Airway resistance is lowest just before noon and rises steadily throughout the rest of the day. This is probably due to the interacting rhythms of plasma adrenaline (which peaks in the early afternoon) and cortisol. Therefore, the assessment of pulmonary function in the morning seems most useful but, by the same argument, asthmatics are likely to suffer most in the evening and during the night.

Another possible application of circadian rhythms to medicine comes from the observation that circadian rhythmicity is often abnormal in illness and that it returns to normal as the patient improves. Even though it is not certain whether the abnormalities are an essential part of the illness or merely reflect drug treatment, it has been argued that an assessment of rhythmicity could aid diagnosis and be an indicator of recovery.

Of much greater potential worth is the suggestion that changed circadian rhythmicity might be a *predictor* of illness. These ideas have been discussed particularly with regard to breast cancers since skin temperature over the affected breast loses its circadian component and instead shows much shorter periods. However, the necessary prospective studies to investigate if changed rhythmicity is a predictor of breast cancer have not yet been performed and, more generally, a convenient rhythm to act as a marker of an illness is by no means always obvious.

6.3.2.2 *Implications for medical personnel.* A knowledge of circadian rhythmicity has implications for medical personnel themselves as well as for the illnesses they treat. Thus, medical teams, in common with other shift workers (see 6.3.5), work less efficiently at night. For example, junior hospital doctors, when played recordings of normal and abnormal heart sounds, are more likely to miss abnormalities at night than during the daytime. One part of the hospital in particular fares badly as a result of these considerations, namely the maternity unit. Twenty-five per cent more births take place at night than in the daytime and so a greater load is placed upon the maternity staff who already have to work during the 'abnormal' night hours.

6.3.2.3 *Drug administration.* Chronopharmacology is potentially a vast area in which knowledge of circadian rhythms can be applied to clinical practice. This new development is based on the finding that the effectiveness of a drug depends upon its time of administration. The effectiveness of the drug (its chronergy) is considered in terms of both the desired (therapeutic) effects and the undesired (toxic) side effects. Any circadian differences might depend upon changes in the uptake, distribution and removal of the drug (its chronopharmacokinetics) or the sensitivity of the subject to the concentration of the drug at its site of action (its chronesthesy). These ideas can be illustrated in man by a consideration of steroid administration and of cancer chemotherapy.

After adrenalectomy, or during adrenal malfunction, synthetic steroids are given. Using a 'chronotherapeutic' approach, two-thirds of the steroid supplement would be given in the morning and the rest later in the day so as to reproduce more closely the circadian rhythm of the steroid (Figure 6.6). However, there are circumstances when other times of administration are believed to be beneficial. The circadian rhythm of respiratory difficulty in asthmatics referred to above is related to the rhythm of endogenous cortisol. In addition, the pain associated with arthritis, immune responses and the effects of allergens are all more marked in the evening and on waking, and have also been linked to the rhythm of adrenal steroids. All these observations indicate possible advantages of giving steroids in the evening. This has been tested, but there are complications. For example, the pituitary gland seems most sensitive to feedback control during the evening and so there is the disadvantage that endogenous steroid secretion is suppressed more with evening treatment. Clearly, more work is needed to decide the optimal timing of steroid administration.

In cancer chemotherapy the nature and dose of the drug are such that the problem is generally of finding a schedule of drug administration which is least toxic to the patient. In principle, the drug should be given at a time when patient susceptibility (drug toxicity) is least. However, more refined possibilities exist, two examples of which can be given. First, one can use the fact that circadian rhythms in cancerous tissues are sometimes out of phase with those in healthy tissues. Therefore, it might be possible to choose a time of administration when the toxic effect of the drug upon the patient is at its lowest but when that upon the cancer cells is not so. Secondly, one can try to reduce the toxicity. One chemotherapeutic agent, *cis*-diamino-dichloroplatinum (DPP), damages the kidney as an undesirable side effect. In animals, such damage can be reduced by simultaneous administration of saline at a time when the kidneys can remove water loads efficiently. In this way the drug is eliminated in a more dilute solution, thus exposing the kidney to a lower drug concentration.

6.3.2.4 *Abnormal clock functions.* Some other clinical conditions are known in which an abnormality of clock function is apparent (but not necessarily the basic cause, of course). For example, in delayed sleep phase syndrome, patients are unable to sleep at normal hours, especially after previously going to bed late, say during the weekend. This might result from the inability of a subject to advance the phasing of his rhythms because of inadequate mechanisms of entrainment and/or an internal clock with a free-running period longer than normal. Whatever the cause, an effective cure is to adjust the patient's sleep time to normal by getting him to retire progressively later each day until he sleeps at a convenient time and then advising him never to stay up beyond this time. Other areas where abnormal rhythmicities are suspected include manic depression and other periodic psychoses. In some cases, they may be associated with a clock that is running with an abnormal period that is less than 24 h; consequently, there is an abnormal (and often changing) relationship between internal and external rhythmicities. Psychotic symptoms seem to be related to the mismatch that exists at any moment. In treating these psychoses it is relevant to note:

(1) that lithium salts (which are often prescribed) are believed to slow the internal clock;

(2) that subjecting the patient to eastward travel (requiring an advance of rhythms, see 6.3.4) temporarily alleviates psychotic symptoms;
(3) that imposing stronger zeitgeber cues (which might more effectively adjust the clock to a 24-h period) is claimed to reduce symptoms.

6.3.3 *Assessing mental performance*

Alone in the Animal Kingdom, man has adopted a lifestyle that can require abnormal sleep–waking routines and, therefore, produces at least temporary dissociation between internal and external rhythms. The two main circumstances where this is so are after rapid time-zone transition and during shift work. In such circumstances a major consideration is the extent to which mental performance is affected. Assessing performance in field conditions such as these is very difficult compared with the much more controllable (but artificial) environment of the laboratory. Some attempts have been made to measure performance at work but such studies are very time-consuming and the data can suffer from many difficulties of interpretation. Suppose a study has shown that the performance of the workforce is less efficient at night than during the daytime. This *might* be due to circadian rhythmicity but some of the other explanations that need to be eliminated first include differences in the nature of the work, in working conditions (lighting, heating, etc.), break facilities (leisure, canteen, etc.), personnel supervision or machine maintenance.

To overcome these difficulties, attempts have been made to produce standardized tests of performance. These tests can vary enormously in complexity and in the kind of performance function they are believed to assess. Table 6.2 lists some of the possibilities, which may be used singly or, better, in groups. Even though such testing sessions can be standardized throughout the 24 h, they are not free of criticism. First, by virtue of being

Table 6.2. Examples of means of assessment of mental performance.

---

(1) Cross out the letters 'E' in a page of prose.
(2) Place dots accurately in randomly arranged circles.
(3) Extinguish a red, but not a green, light (by pressing a switch) when they are flashed on in random order.
(4) Assess the truth, or otherwise, of statements like 'A does not follow B in the sequence BA'.
(5) Sort playing cards into suits (or colours, etc.).
(6) Perform mental arithmetic.
(7) Search for particular sequences (e.g. BX or DJBHIO) in strings of random letters.
(8) Transcribe symbols into letters in coded sequences.
(9) Indicate whenever a particular signal (e.g. length of note) differs from a normal value.
(10) Copy irregular figures seen in a mirror.

---

standardized, they are repetitive and boring to perform. Also they are affected by the conditions (lighting, noise) under which the tests are performed and are time-consuming, often an unacceptable drawback as far as the employer is concerned. Finally, and perhaps most importantly, there is uncertainty as to how well such groups of tests mimic the field tasks. Even so, as with field tasks, subjects perform such tasks less well at night than during the daytime.

These difficulties, coupled with technological advances that allow automated measurement (e.g. of rectal temperature or blood pressure) or analysis of small samples of blood or urine (e.g. for hormones), have led to the search for more convenient alternatives. Briefly, two rhythms that seem to be useful in this regard are deep body temperature and urinary adrenaline concentration. Recording oral temperature or collecting urine samples does not intrude unacceptably upon a subject's activity and such procedures are less affected by external interference than are the tests of Table 6.2. Figure 6.4 predicts that a direct relationship exists between deep body temperature, adrenaline concentration and performance rhythms. This is generally, but not invariably, the case. For example, there is often a post-lunch dip in performance which is not mirrored in the temperature rhythm. Also, there is evidence that short-term memory is actually better at lower body temperatures; thus tasks that involve short-term memory to any major extent (see the second example in test 7 of Table 6.2) might be expected to show a rhythmicity phased differently from that of temperature.

In summary, field performance can be very difficult to measure reliably and so laboratory-based tests of mental performance have been devised. These are often time-consuming or require unacceptably stringent conditions in which the tests are to be performed. As a result, rhythms that are less labile and more convenient to measure (e.g. urine adrenaline concentration, body temperature) have been used. In the account that follows, reference will be made generally to these marker rhythms; both their usefulness and the assumptions that underlie their use must be remembered.

6.3.4 *Time-zone transition*
When a person flies across a number of time zones, there is an immediate mismatch between the new local time and 'body time' as indicated by his internal clock; the very stability which is naturally conferred upon an individual's rhythmicity means that the internal clock is slow to adjust to changes in schedule. As a result, he feels tired, wants to sleep at the wrong time (night-time in the old time zone) and is unable to sleep at the new night-time. His appetite and gastrointestinal functions are also adversely affected and he may feel irritable. This group of symptoms is called 'jet-lag syndrome'. It is not due directly to anxiety over flight, loss of sleep or

change of environment since it is not present after long flights to the north or south.

Studies of circadian rhythmicity suggest that inappropriately phased rhythms are induced temporarily after transmeridianal flights. Figure 6.7 shows the temperature rhythm of eight subjects after two flights across six

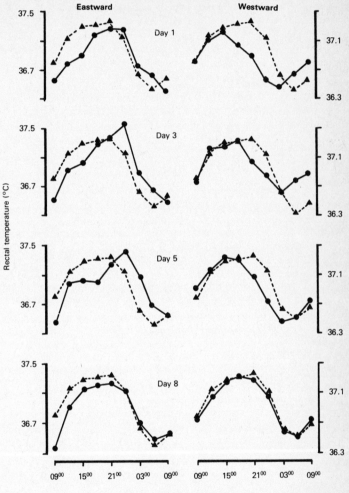

Fig. 6.7 The effect of transmeridianal flight through six time zones in an eastward or westward direction upon the rhythm of rectal temperature. ▲---▲, pre-flight data, represented for clarity on each post-flight day; ●——●, post-flight data. Data represent the means from eight subjects. From Klein, K.E., Wegmann, H.M. & Hunt, B.I. (1972). *Aerospace Medicine* **43**, 119–32.

time zones, one to the east and the other westwards. After the eastward flight, adjustment remains poor up to day 5. Thus, the pre- and post-flight body temperatures do not coincide when expressed on local time as would be the case had adjustment been complete. Instead, temperature rhythms appear by new local time to be delayed, which is explicable if it is remembered that the new local time is ahead of the old after an eastward shift. By day 8, adjustment is essentially complete, the phases of the two curves (but not their absolute values) being the same. Adjustment takes place because the traveller is exposed to differently phased zeitgebers appropriate to the new time zone. Thus, if travellers are sheltered from some of these zeitgebers (by remaining in hotel rooms instead of mingling with the new society), then adjustment is slower.

Figure 6.7 also shows that adjustment to westward travel occurs more quickly (the two curves coincide at least partially by days 3–5). This is a general finding and may be explained as follows. After an eastward shift, the internal clock lags behind new local time and so has to adjust by advancing its phase; after a westward shift, adjustment of the clock will be by delay. Since the free-running period of the internal clock is greater than 24 h (see 6.2.1), adjustment to a westward shift (involving a delay in the phase of rhythms) is easier. A corollary of this is that animals that possess clocks with free-running periods less than 24 h (e.g. finches) adjust more rapidly to an eastward time-zone shift. The position is slightly more complicated, however, since not all parameters adjust to a time-zone shift at the same rate. In general, the larger the exogenous component of a rhythm, the more rapidly it appears to adjust. Inspection of Figure 6.4 suggests, therefore, that rhythms of temperature and plasma adrenaline concentration will adjust less quickly than those of performance. That is, these two rhythms give some indication of the course of adjustment of the performance rhythms, but they will not mirror them exactly.

In summary, the individual who is suffering from 'jet-lag syndrome' will show the following abnormalities in his rhythms. First, the rhythms will be out of phase with the new local time; secondly, they will be adjusting; and thirdly, they will be adjusting at different rates and so will be out of phase with each other. Which of these abnormalities, if any, is responsible for 'jet-lag syndrome' is unknown.

Can one offer advice to the traveller? The advice given depends upon how long he intends to stay in the new time zone. If the visit is to be a short one, adjustment of the internal clock does not take place; all the traveller can do is to arrange appointment times that are coincident with both old and new daytime (for example, the new evening after eastward travel and the new morning after a journey to the west). Meetings in the morning after an eastward flight (or evening after westward) are undesirable since they coincide with the old night-time. If the traveller is to remain in the new time zone, he should attempt to adjust rapidly by exposing himself as fully as possible to the new zeitgebers. The time required for complete adaptation is difficult to predict; there is some evidence that data such as those given in Figure 6.7 underestimate the time required for the internal clock to adjust. This is because the rhythms have been measured in the presence of rhythmic exogenous factors that are fully adjusted to the new time zone (see 6.2.3). Even so, adaptation is generally accepted as being complete after about 10 days. It might be possible to 'pre-adapt' somewhat by adopting the sleep–waking pattern of the new environment for some days before the flight. However, business and social constraints render this rarely possible and it is not a very effective method anyway! This is because it resembles shift working, to which we now turn.

6.3.5 *Shift working*
When people work at night, they are required to be alert when they wish to sleep and to sleep when they should be preparing for a new day. People who work at night often complain of symptoms similar to those of 'jet-lag syndrome' and the cause is believed to be similar. With continued night work, some workers find the inconvenience and malaise unacceptable and so leave night work permanently. As a group, shift workers suffer more from gastrointestinal disorders than the population as a whole. It would be unwarranted to attribute these problems directly to the abnormal sleep–waking schedule. For example, workers at night tend to eat snacks rather than a single large meal, to smoke more, and to drink more alcohol, any of which might be responsible for some of the disorders. Nevertheless, abnormal circadian rhythms are suspected to play some role in the general malaise suffered by many shift workers.

6.3.5.1 *Altered rhythms in shift workers.*
The shift worker sleeps less well during the daytime than during the night. This is not wholly due to light and noise, for it is found also in the controlled environment of the laboratory-based experiment. It is due to the internal clock which, by the end of the night, is preparing the body for a new day. If, in spite of a rising deep body temperature and blood adrenaline concentration, the shift worker does sleep, then increasing urine production and the need to micturate are likely to wake him. Not only the quantity, but also the quality of sleep are changed in the daytime in two ways. First, the distribution of the different sleep stages within the sleep period is altered and, secondly, the timing within sleep of hormone secretion is modified. Thus, in normal sleep, most stage 4 sleep (deep sleep) occurs at the beginning, coincident with growth hormone secretion, and most rapid eye movement (REM) sleep is coincident with cortisol secretion at the end. During daytime sleep, growth hormone

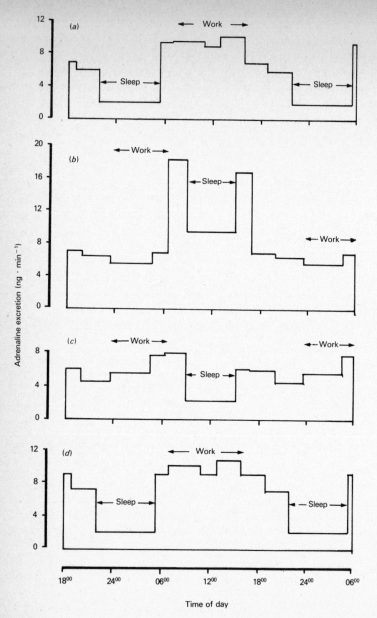

Fig. 6.8 Daily course of urinary adrenaline excretion in shift workers during: (*a*) the last week of day work; (*b*) the first week of night work; (*c*) the third week of night work; (*d*) the first week of return to day work. Data are means from 15 workers. For clarity data have been repeated beyond 24 h. From Akerstedt, T. & Froberg, J.E. (1975). In: *Experimental Studies in Shiftwork*, eds P. Colquhoun, S. Folkard & P. Knauth, pp. 78–93. Westdeutcher Verlag: Opladen.

secretion still mainly occurs at the beginning (i.e. it adjusts rapidly) as now does most REM sleep (which adjusts slowly). Stage 4 sleep tends to be distributed throughout sleep, and cortisol rhythms, having a large endogenous component, are slow to adjust. Any implications that these abnormalities might have for health are not known, but to change the normal relationship between sleep stages and hormone secretion might have unpleasant side effects.

When other rhythms are considered, they have in common a slow adjustment to night work (Figure 6.8). Even after 21 continuous nights of work, urinary adrenaline excretion does not reach normal 'working' values (during the night-time) even though the values associated with sleep (during the daytime) do seem to adjust. Similar incomplete adjustments with oral temperature have been observed. This difficulty of poor adjustment is compounded by the fact that loss of adjustment during days off (e.g. at the weekend) or during day shifts is alarmingly rapid.

Why is this so? It is probably due to the zeitgebers acting upon shift workers. With time-zone transition all zeitgebers unambiguously signal new local time; with shift work there is ambiguity. Thus, although the worker can adjust his lighting, sleep–waking pattern and even mealtimes to fit in with his work, he is aware that many other social factors have not changed. This will be obvious to him during his mid-work break, when he wants to watch television, or during his leisure time when he would like to be involved in social events. By contrast, when he reverts to normal routines during weekends, all his zeitgebers are in accord and so a rapid adjustment of his rhythms to his normal lifestyle takes place.

Thus the conventional weekly shift system in Britain will show poor adjustment during the week, loss of adjustment at the weekend and the need for the process to have to begin all over again when night work restarts.

Because of these problems, alternative shift systems have been suggested. One that is popular in much of Europe is a rapid rotation of shifts whereby each shift is only worked once or twice in succession before moving on to the next in the sequence (the sequence being morning–evening–night). Certainly, such rapid changes in the sleep–activity pattern are too rapid for any useful adjustment of rhythms to take place and its proponents argue that this is one of its merits.

Rapid rotation of shifts reduces the adjustment of circadian rhythms, but this also means large mismatches between internal and external rhythms whenever night work is performed. By contrast, as shown in Figure 6.8, a slow rotation of shifts is associated with less stable rhythms but at least partial adjustment to night work can occur. Until the cause(s) of 'jet-lag syndrome' and shift workers' malaise are known (see 6.3.4), a choice between shift systems, aimed at reducing these undesirable side effects, cannot logically be made.

6.3.5.2 *Advice to shift workers.* Can one, even so, offer advice to the shift worker and those who devise shift systems? A comment can be made relating to the direction in which successive shifts change. The sequence that is normally found (morning shift–evening shift–night shift) requires habits to become later as the sequence progressess; interestingly, the preferred direction of travel for circumglobal air crews (i.e. westwards) has the same requirements. In both cases, a lengthening of the day is easier for a clock with a natural period in excess of 24 h.

A study of experienced shift workers or transmeridianal travellers may give clues as to how the undesirable effects can be minimized. (However, it must be realized that such people probably do not represent a random sample of the population, having been self-selected on the basis of showing few unpleasant symptoms.) Experienced shift workers and time-zone travellers do seem to be slightly better adjusted to their schedule than less experienced personnel. In part, this may reflect their acceptance of what shift work implies, inasmuch as they sometimes plan their social life around their shift work. Alternatively, it is possible that rhythms adjust more rapidly if changing schedules are experienced for long enough, but there is no good evidence for this. Another possibility exists, namely that some individuals are naturally better suited to shift work or continual time-zone transitions than are others. Recently, considerable interest has been shown in the possibility of selecting individuals who will be affected least adversely by changes in routine. Those who have been suggested are:

(1) Individuals who possess rhythms with an amplitude that is higher than usual. It is suggested that such rhythms adjust less following changes in routine.

(2) Individuals who possess rhythms that are phased later or earlier than usual. Rhythms that are phased unusually late have been suggested to aid work at night and sleeping afterwards during the morning. Rhythms phased earlier might be an asset to people having to retire and rise early for morning shifts.

(3) Individuals of appropriate temperament. The person least affected by shift work would be the worker who is prepared to adjust his hours of sleep to fit in with his work and who can return to sleep after having been woken prematurely by the rest of society.

In all cases, evidence is sparse and further studies are required.

A few words of caution here. Some of the difficulties that arise after time-zone transition and during shift work have been described. They are based upon the abnormal timing of rhythms that is produced. However, the decision of an individual whether or not to continue shift work or to continue crossing time zones will be based upon many factors in addition to the physiological ones. Further, if tests could be devised that reliably indicated individuals who would find shift work or time-zone transitions difficult to adjust to, then complex issues are raised that are not within the expertise of the physiologist.

6.4 Further reading

6.4.1 *Elementary*

Mills, J.N., Minors, D.S. & Waterhouse, J.M. (1981). Biological rhythms. In: *The Principles and Practice of Human Physiology*, eds O.G. Edholm & J.S. Weiner, Chapter 9. Academic Press: London, New York.

Palmer, J.D. (1976). *An Introduction to Biological Rhythms*. Academic Press: London, New York.

Saunders, D.S. (1977). *An Introduction to Biological Rhythms*. Blackie: Glasgow.

6.4.2 *More advanced*

Wever, R. (1979). *The Circadian System of Man*. Springer-Verlag: Berlin.

Minors, D.S. & Waterhouse, J.M. (1981). *Circadian Rhythms and the Human*. John Wright: Bristol.

Moore-Ede, M.C., Sulzman, F.M. & Fuller, C.A. (1982). *The Clocks that Time Us*. Harvard: Cambridge, Mass.

Chapter 7

Changed acceleration forces (gravity)

Summary

The immediate effects of gravitational forces on the body are exerted mainly upon the cardiovascular system. With increased acceleration in the headwards direction, the normal gradients of pressure seen in upright man are accentuated. This causes a decreased blood flow to the eyes (causing limited, or loss of, vision) and to the brain (unconsciousness). In a zero-G (weightless) environment, there is a movement of fluid towards the thorax since normal pressure gradients have been lost. If weightlessness persists for some days, diuresis will correct such engorgement. However, other longer-term changes persist over the course of weeks and months: the decreased workloads required of the circulation, muscles and bones lead to a general 'deconditioning'; cardiac and skeletal muscle atrophy and bones lose calcium. Even though these changes can be viewed as an acclimatization to weightlessness, they pose serious problems when a return to Earth is considered. Attempts to minimize such deconditioning — by exercise and by pressure suits designed to pool blood in the periphery — have been only partially successful.

7.1 Forces acting upon the human body

Gravity draws us towards the centre of the Earth with a force that varies negligibly from the top of the highest mountain to the bottom of the deepest mine. From one of Newton's Laws of Motion, any movement, e.g. ascent in a lift or the 'spring' in a person's gait, will require a force to be exerted upon the body. These forces are transient and small, but the development of fast means of transport has increased them. In normal journeys by car, the forces due to acceleration, braking and cornering are rarely greater than 1-*G* (that is, equal to the force exerted by gravity) but they can act for several seconds. In commercial flights, the forces experienced result mainly from circling manoeuvres and are not expected to be greater than 1.3-*G*; but for test pilots and those involved in aerial displays or combat, transient forces up to 8-*G* are not uncommon. Generally these forces act in a headwards direction — as when a plane is turning and the head of the pilot is tilting towards the centre of rotation, but they can act in the opposite direction — as when the pilot is flying 'upside-down' or circling with his head pointing away from the centre of rotation. The advent of rocket travel has required acceleration forces to be experienced for minutes rather than seconds, both during the launch and during re-entry into the Earth's atmosphere after a flight.

One way to investigate the effects of acceleration forces upon the body is in a plane flying at the appropriate speed in a circle of the appropriate radius. A cheaper and more flexible alternative, which allows experiments to be performed in laboratory conditions and enables greater forces to be experienced, is the centrifuge (Figure 7.1). With this apparatus, the force exerted can be determined from the radius of the turning circle and the speed of rotation. Even though the Earth's gravitational force continues to act vertically, the direction in which the imposed centripetal force is applied with respect to the subject can depend upon his posture. Most work has been concerned with the effect of headwards acceleration (in which the subject's head tilts towards the centre of rotation of the centrifuge); much less work has been done upon accelerations applied in other directions.

A special case exists when the body is falling freely, that is, when the forces acting upon the body are not being opposed so that the condition of zero-*G* or 'weightlessness' is experienced. This occurs in satellites orbiting around the Earth and can exist for weeks or more, rather than seconds or minutes as in the case of forces producing movement. Hence, acclimatization to prolonged weightlessness requires consideration, as do the problems that arise when a return to Earth takes place. Special techniques are required for a study of weightlessness: these will be dealt with later.

As we shall see, many of the physiological problems caused by accelera-

Fig. 7.1 The human centrifuge at the RAF Institute of Aviation Medicine. It has a radius of 30 ft and can impose accelerations of up to 30-*G* upon subjects in the cabins at the ends of the arms. Reproduced by kind permission of Air Commodore P. Howard.

tion arise because of the presence in the body of columns of fluid. In any such column, there will be a difference in pressure, *P*, between two points given by the equation:

$$P = h \times \varphi \times G$$

where *P* can be conveniently expressed as cmH$_2$O, *h* is the difference in vertical height (cm), φ is the density of the fluid (g·cm^{-3}) and *G* is the acceleration force acting upon the column (expressed as a multiple of the normal gravitational force).

As this equation indicates, the pressure gradient due to differences in vertical height varies in size and direction in accordance with the acceleration forces acting upon the body and will disappear in the weightless condition.

7.2 Physiological effects of acceleration

7.2.1 *The central nervous system*
If a subject is suddenly accelerated in a headwards direction with a force of about 4-G, peripheral vision is impaired and perception of colour and detail by the fovea deteriorates. These symptoms are generally called 'grey-out'. With increased acceleration (4.5 to 5-G), vision, but not consciousness, is lost (black-out). In both cases, the symptoms often disappear when acceleration is maintained but, with faster accelerations, unconsciousness is produced. If acceleration is gradual, symptoms of visual malfunction are often not experienced: instead there is, without warning, a sudden loss of consciousness.

These visual defects result from decreased retinal blood flow. The decrease in flow is due to a fall in pressure across blood vessel walls, which thereby decreases vessel calibre and raises vascular resistance. This decrease in transmural pressure can be deduced as follows:

(1) Mean arterial pressure at the level of the heart is 125 cmH$_2$O.
(2) The eyes are above the level of the heart by 25 cm.
(3) Intraocular pressure is 20 cmH$_2$O.

Therefore in a 1-G environment, assuming a density for blood of 1 g cm^{-3}, the transmural pressure responsible for maintaining the patency of the arteries is:
$$125 - (1 \times 25) - 20 = 80 \text{ cmH}_2\text{O}$$
However, in a 3-G environment, this transmural pressure declines considerably to:
$$125 - (3 \times 25) - 20 = 30 \text{ cmH}_2\text{O}.$$
By contrast, the arteriovenous gradient (responsible for causing perfusion of the retina) does not change since arterial and venous pressures change equally.

Persistent acceleration stimulates the baroreceptor reflexes, since the carotid sinus receptors are also above the heart, and so raises blood pressure and retinal blood flow. If acceleration is slow, the baroreceptor reflexes are recruited slowly, as during a prolonged haemorrhage, and sympathetic tone rises progressively. If this circumstance persists too long, the ability of the baroreceptor reflexes to compensate is exceeded: vaso-vagal syncope (faint) ensues, there being a marked increase in vagal tone, a bradycardia and peripheral vasodilatation.

In some studies, acceleration *downwards* (i.e. away from the head) has been found to produce 'red-out', in which vision is blurred and appears to be through a red filter. Damage to retinal blood vessels has not been observed, but burst vessels in the conjunctiva have been described and may explain the phenomenon. The wearing of goggles which raise the pressure outside the eye is an effective countermeasure.

Brain function in general seems less susceptible than vision to the effects of acceleration (for example, consciousness is maintained during 'blackout'). This is because the perfusion of the brain vascular bed is preserved, owing to a number of factors:

(1) The brain is surrounded by an indistensible cranium and its centre is filled with fluid (the cerebrospinal fluid). Therefore, changes in pressure inside blood vessels produced by acceleration are reflected very closely by changes in cerebrospinal fluid pressure. Further, since brain interstitial fluid, cerebrospinal fluid and perivascular fluid

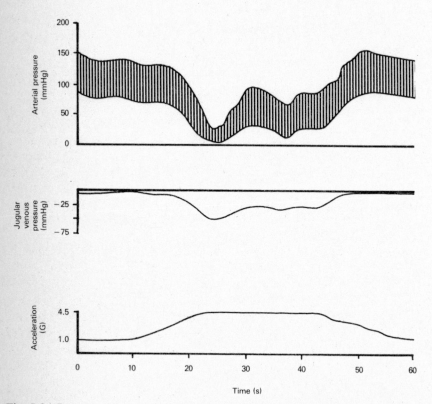

Fig. 7.2   Pressures in the radial artery (supported at eye level) and in the jugular bulb during exposure to 4.5-*G* in a headwards direction. The subatmospheric pressure in the vein acts as one limb of a siphon, to sustain cerebral blood flow. From Howard, P. (1982). Acceleration. In: *The Principles and Practice of Human Physiology*, eds O.G. Edholm & J.S. Weiner, Chapter 4. Academic Press: London, New York. Reproduced by permission © Academic Press Inc. (London) Ltd.

pressures are equal, little change in blood vessel transmural pressure is observed. The eyes, of course, are outside the cranium.

(2) As with the eye, blood flow through the tissue is determined also by the arteriovenous difference in pressure. During acceleration, venous pressure is reduced by the same amount as arterial pressure; the driving force is therefore maintained. Some claim that the jugular vein is, in some way, protected from collapse when intraluminal pressures decrease.

(3) If brain perfusion does decrease, autoregulation of vascular tone occurs (initiated by local chemical changes) and is very marked.

Therefore, provided that the vessels do not collapse, blood flow to the brain is maintained surprisingly well (Figure 7.2), though deterioration in mental performance before unconsciousness occurs has been measured in some studies. If the blood vessels collapse, or if, for some ill-understood reason, the vaso-vagal reflex is initiated, unconsciousness occurs very rapidly.

7.2.2 *The cardiovascular and respiratory systems*

An immediate consequence of headwards acceleration is a decrease in cardiac output. This can be explained as follows. Acceleration causes the internal organs of the body to move away from the head: this reduces intra-thoracic pressure. Even though the pulse pressure developed during cardiac systole remains the same, it is taking place against a reduction in intrathoracic pressure and so there is a net fall in driving force. Other early changes include an increased pooling of blood in the dependent parts of the body, with a consequent decreased venous return. Both these factors lower cardiac output and, together with an additional fall in pressure at the level of the carotid sinus (because it is above the heart), cause the baroreceptor reflexes to be stimulated in an attempt to maintain an adequate blood pressure.

In terms of respiration, acceleration affects both breathing movements and gas exchange in the lungs, the latter due to changed ventilation: perfusion ratios at different levels of the lungs. (For an explanation of these ratios, the reader is referred to a standard textbook.) With headwards acceleration, the movement of internal organs facilitates inspiration but hinders expiration. Not surprisingly, tidal volume tends to encroach upon the inspiratory reserve. Acceleration away from the head produces the opposite effects. As we shall see later, many of the undesirable effects of acceleration can be minimized by altering body posture so that acceleration is in a forwards direction: however, in this orientation, the outward movements of the chest wall necessary for inspiration are hampered.

The normal lung shows gradients of blood perfusion and alveolar ventilation from its apex to its base. Thus, compared with an 'average' alveolus, the apical alveoli are more fully distended, underventilated and underperfused and the basal alveoli are less distended, overventilated and overperfused. Since these gradients of perfusion and ventilation are produced by factors which are affected by acceleration forces, they require consideration here. Acceleration in a headwards direction accentuates the gradients whereas acceleration in the opposite direction can reverse them. During headwards acceleration, the apical alveoli might be almost fully distended throughout the respiratory cycle but receive very little blood or ventilation. By contrast, the basal parts of the lung receive large amounts of blood (so much so that there is even the possibility of pulmonary oedema), but ventilation does not increase proportionally. Further, some airways might collapse, thereby trapping alveolar gas, as might some alveoli (atelectasis).

As a result of these changes, the apex and base of the lung function poorly with respect to gas exchange. Cardiac and ventilatory work is wasted (venous shunts and alveolar deadspace, respectively) and hypoxaemia develops. Difficulties in ventilation and thoracic pain may occur and are believed to be associated with some of these disturbances. Such symptoms are relieved by administering pure oxygen, but this procedure makes atelectasis (alveolar collapse) more likely. Thus, if airways collapse while the alveoli contain oxygen alone, continued gas exchange will result in alveolar collapse. (As alveolar volume decreases, the PCO_2 rises so that CO_2 also will leave these alveoli.) By contrast, if nitrogen is present as well, then this gas prevents collapse of the alveoli and CO_2 is exchanged for oxygen.

7.2.3 *Countermeasures*

The effects of acceleration can be minimized by appropriate orientation of the body. If it is possible for the subject to accelerate 'forwards', pressure gradients will be lessened because the thickness of the body is less than its height. Obviously, some degree of compromise might be required if the subject has tasks to perform, but a greater tolerance to G-forces has been found when the body is positioned in this way. Other behavioural changes involve curling up the body, tensing the abdomen or pulling up the legs. Also transient Valsalva manoeuvres (i.e. forced expiration against a closed glottis) or shouting have been used. All these procedures increase venous return and raise intra-abdominal pressure. This latter factor prevents the fall of the diaphragm, so maintaining intrathoracic and aortic systolic pressures.

An artificial aid that has been developed is the 'anti-gravity suit'. It is based upon the principle that water immersion would counter the gravitational effects already described by producing an equal gradient outside the body. Recent designs have been air-filled rather than water-filled and have tended not to encase the lower body completely. Anti-gravity suits are believed to work for three main reasons: they prevent the descent of the diaphragm; they reduce the pooling of blood in the legs; and they ensure that any blood expelled from the mesenteric vessels is moved towards the heart. The parallels between these aims and many of those achieved by behavioural changes already described are obvious. Recent work has shown that combinations of countermeasures are more effective than when the measures are used separately.

7.3 Physiological effects of weightlessness

If man is to explore the solar system by his present means of travel, he will have to endure weightlessness for years. Can man adjust to such periods of zero-*G* and, if so, what happens when he returns to live again in a gravitational field?

7.3.1 *Means of study*

Before man's first space flight it was necessary to have some idea of how he would react to weightlessness over a period of at least a few days. Earthbound investigators used constant bed-rest and water-immersion techniques to mimic the effects of weightlessness. In complete bed-rest, gravity still exists, but its effects are no longer exerted along the length of the body. Studies lasting up to 190 days have been made. A major difficulty with this method is in deciding the extent to which changes in posture should be allowed for meals and toilet and for practising tasks that might be required in an orbiting satellite. Water-immersion studies, the rationale for which is identical to that for the 'anti-gravity suit' already discussed, suffer from different problems. Thus, the temperature of the water affects considerably the subjects' comfort and ability to sweat; subjects have to be separated from the water by an impervious sheet to prevent damage to the skin; and movement is difficult because of the viscosity of the water.

In spite of these problems, data from both types of study are similar to those obtained so far in the prolonged space flights (up to 190 days) of the 'Soyuz', 'Salyut', 'Skylab' and 'Spacelab' programmes. For this reason, the results from all techniques will be considered together in the following account.

7.3.2 *Body fluids*

During weightlessness, there is a redistribution of about 2 litres of body fluids away from the dependent parts of the body and towards the head, neck and thorax. For about 3 days the head and neck feel bloated and the neck veins are engorged whereas the legs become thinner. During this period, a diuresis takes place, probably via changes in anti-diuretic hormone (ADH) secretion, initiated by volume receptors in the atria and great veins (Gauer–Henry reflex). This diuresis corrects the excess of fluid in the thoracic veins and so reduces the increased cardiac output and load on the heart due to excess venous return. In space flight, about 1.5–2 litres of fluid are lost in total and plasma volume falls by about 10 per cent. Water-immersion studies produce larger changes. Two mechanisms probably contribute to this difference. First, astronauts decrease water intake during a flight: secondly, water immersion compresses the interstitium.

Weightlessness causes the red cell count to fall. Since red cell breakdown is believed not to change, this implies that erythropoiesis falls by more than the decrease in plasma volume that has just been described. Indeed, bone marrow production almost ceases, at least during the first days of weightlessness.

7.3.3 *Cardiovascular system*
Initially, weightlessness causes an increase in cardiac output due to venous engorgement. After correction of this engorgement (by day 4), cardiac output and heart rate fall. These changes are considered to reflect decreased sympathetic tone and increased parasympathetic tone associated with the decreased work load of the heart. In apparent confirmation of this hypothesis, end-diastolic volume increases. Some reports suggest that, with longer periods of weightlessness (over 50 days), a degree of ventricular atrophy occurs. This general 'deconditioning' of the cardiovascular system under prolonged zero-*G* conditions can be seen also when the responses to circulatory 'stresses' are considered. Thus, the cardiovascular changes produced by exercise are more marked and, after experimentally decreasing venous return (e.g. by inflating cuffs around the thighs), the time that elapses before fainting occurs is decreased.

7.3.4 *The central nervous system and movement*
A substantial proportion of astronauts experience giddiness, disorientation, nausea and even vomiting during the first few days in space. These symptoms are not observed during studies performed on Earth (thus showing the problems associated with extrapolation of data from indirect studies of weightlessness) and are attributed to an imbalance between different neural inputs to the brain. Thus, the input from the semicircular canals (which reflects angular acceleration) is similar to that found on Earth, but inputs from skin, muscle and joint receptors (which signal the postural forces acting on the body) and the otolith receptors (which signal the orientation of the head in a gravitational field) change considerably.

Movement is limited, which is sometimes a problem (see Chapter 8), and there is far less need for tonic activity in 'anti-gravity' extensor and flexor muscles. As a result, muscular atrophy occurs as indicated by direct measures of strength and fatigue and by the increased excretion in the urine of urea.

Some studies have suggested that postural reflexes become exaggerated with continued weightlessness. The 'resetting' that this implies is similar to that described previously for the cardiovascular system in response to exercise and decreased venous return.

7.3.5 *The skeleton*

All the changes so far described reach equilibrium after a few weeks or months, when acclimatization to weightlessness might be thought to be complete. However, demineralization of bones continues at an unchanged rate during 6 months of bed-rest or during space flights of the same duration (which is the longest time that the process has been studied). This loss of bone material can be measured directly from densitometric studies of individual bones or indirectly from rates of urinary excretion of hydroxy-proline and calcium. The bones most affected are those that normally bear the weight of the body. The rate of loss of minerals is about 0.5 per cent of the total per month: this implies a considerable danger of fractures on return to a normal, 1-*G* environment after about 2 years in space. The increased urinary calcium concentration might also increase the incidence of renal calculi.

7.3.6 *The return to Earth*

From the above we can conclude that, to a very considerable extent, man can acclimatize to a zero-*G* environment. The problem which remains is that of being able to withstand a gravitational field on return to Earth (or any other planet). The potential hazard from bone demineralization has been mentioned already, but other problems exist also. On return to terra firma, blood once again pools in the dependent parts of the body (Figure 7.3). Bearing in mind that the blood volume is decreased by at least 10 per cent, that tissue tone might be reduced and that a further 'resetting' of the vestibular reflexes is necessary (since fresh inputs from the otoliths and receptors involved in kinaesthesia are being sent to the central nervous system), the difficulties of standing upright become obvious. A decreased tolerance of the cardiovascular system to the stresses imposed by decreased venous return and exercise (found also during prolonged weightlessness, see 7.3.3) is observed immediately after the return to Earth.

In the days following their return, astronauts drink copiously, their kidneys conserve salt and the number of circulating reticulocytes increases as the anaemia is corrected. More slowly, cardiac muscle, skeletal muscle and bone all recover. The exact rate of recovery has not been studied at all fully in ground-based simulations of weightlessness: a single astronaut who

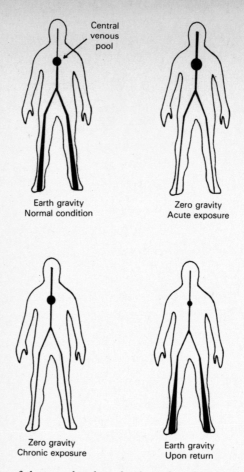

Earth gravity
Normal condition

Zero gravity
Acute exposure

Zero gravity
Chronic exposure

Earth gravity
Upon return

Fig. 7.3   Diagram of the postulated mechanism of cardiovascular 'deconditioning' explained in the text. From Howard, P. (1982). Acceleration. In: The *Principles and Practice of Human Physiology*, eds O.G. Edholm & J.S. Weiner, Chapter 4. Academic Press: London, New York. Reproduced by permission © Academic Press Inc. (London) Ltd.

underwent a second flight 6 months after the termination of a prolonged first flight seemed completely recovered.

7.3.7 *Countermeasures*
Much effort has gone into minimizing the unpleasant side effects and the 'deconditioning' which are associated with prolonged weightlessness. Thus, since nausea is most marked following head movement, some Russian astronauts have worn caps that restrict such movement. Methods aimed at opposing the process of 'deconditioning' have attempted to reproduce the

normal 'stresses' that occur in a 1-G environment. Thus, astronauts have been subjected to techniques that cause blood to be pooled in the legs and so decrease venous return; work schedules with a bicycle ergometer, springs or isometric contractions have been tried; and 'constant-load suits', which require the continuous expenditure of muscle effort, have been worn. Reports of recent space flights have emphasized such work schedules. Just before the return to Earth, further measures have been used. For example, astronauts have worn 'anti-gravity suits' and have had their fluid volume deliberately expanded by drinking saline. Post-flight investigations indicate that these measures reduce the cardiovascular 'deconditioning' produced by space flight and decrease the problems after returning to Earth. However, the changes in erythropoiesis, vestibular function, skeletal muscle, cardiac muscle and bone wasting are by no means completely prevented. Increasing the duration and intensity of exercise becomes impracticable since it intrudes upon the astronauts' duties; available space and the astronauts' motivation are also limiting factors. A basic problem is that, under terrestrial conditions, gravity exerts its effects continuously. To duplicate this in space would require the creation of an artificial gravity by imparting a spin (and hence centrifugal force) to the spacecraft. The size of centrifugal force needed to prevent 'deconditioning' is not known. Since the centrifugal force is proportional to the radius of rotation as well as its speed, if the force required to prevent 'deconditioning' is at all substantial, then the angular velocity of a small spacecraft (but not a large space-station) might be unacceptably large to allow for navigation and routine manoeuvres.

7.4 Further reading

European Space Agency Publications. ESA BR-01 (1979). *Biology and Medicine in Space*, ed. H. Bjurstedt. ESA BR-09 (1982). *Microgravity as an Additional Tool for Research in Human* Physiology, eds F. Bonde-Peterson, H. Hinghofer-Szalkay & J. Hordinsky.

Howard, P. (1982). Acceleration. In: *The Principles and Practice of Human Physiology*, eds O.G. Edholm & J.S. Weiner, Chapter 4. Academic Press: London, New York.

Medical Report on Skylab (1976). *Aviation, Space and Environmental Medicine* **47**, 347–424.

Chapter 8

Self-contained environments

Summary

Technology has enabled man to live in otherwise uninhabitable environments such as underwater or in space. However, for man to function normally, several environmental parameters have to be controlled, most notably the composition of the atmosphere and temperature. There are also many psychological problems to be overcome. Some of these are outlined in the present chapter.

8.1 Introduction

Man's inquisitiveness and inventiveness have led him to explore all parts of the Earth and eventually to conquer hostile environments that are incapable of supporting life. This conquest has been made possible not by physiological adaptation but by engineering techniques that provide man with a closed environment whose composition can be closely controlled. Such environments are required during travel in aircraft at high altitudes, space flights and submarine voyages. Some babies have rather similar requirements, particularly if they are born prematurely and are ill, when they cannot exist in a normal environment. The philosophy has been that if man cannot survive in a particular environment, then a suitable artificial one must be provided.

In these circumstances physiologists usually only specify the environmental limits within which engineers have to work. As engineering techniques improve, these specifications can more nearly approach 'normal' values. This is obviously most important when long-term confinements are contemplated, as man can tolerate wider limits for shorter periods. The most important physiological considerations apply to the composition of the artificial atmosphere, the temperature, diet and exercise. Many psychological factors, such as recreation, work–rest schedules, personal hygiene and social interactions, must also be considered. Not all these considerations, of course, apply to all situations where man lives in an artificial environment.

8.2 The artificial atmosphere

The range of PO_2 in which humans can live in the short term is discussed elsewhere (see 4.2.1). These limits are important in commercial flights. At a cruising altitude of 10 000–12 000 m the atmosphere cannot support life. The aircraft fuselage is therefore pressurized. Engineering and economic constraints prevent pressurization to the value at sea level and a value equivalent to an altitude of 2000 m is chosen, where the PO_2 is about 105 mmHg (14 kPa) compared with 160 mmHg (21 kPa) at sea level. Most normal, fit people can exist without problems at this PO_2, though some patients with respiratory or cardiovascular disorders may find themselves in difficulties.

In the longer term the limits that are believed not to cause detriment to the manipulative and cerebral skills required by astronauts and submariners are much narrower. For example, an inspired PO_2 of less than 130 mmHg (17 kPa) adversely affects night vision. In modern submarines the atmospheric PO_2 is maintained at 150 mmHg (20 kPa). High PO_2 values may also be detrimental.

The partial pressure of carbon dioxide is also important. If inspired PCO_2 reaches 22 mmHg (3 kPa), ventilation rises, reflexes slow down and general mental efficiency declines. Subjects complain of headache and marked fatigue with only slight exertion. Inspired PCO_2 must be kept below 11 mmHg (1.5 kPa) to sustain normal performance and recent research suggests that lower levels are preferable. Any rise in inspired PCO_2 is similar in effect to a mild respiratory acidosis and changes in ventilation and bicarbonate handling by the kidneys and the choroid plexus can be expected to occur. The changes in calcium metabolism in bone appear to be complex. Thus, with inspired PCO_2 greater than 1.5 kPa, there is evidence that demineralization of bone occurs and calcium homeostasis is disturbed, but with smaller changes in inspired PCO_2, there seem to be cyclical changes in calcium metabolism. At first, urine calcium is low and the bone accumulates bicarbonate; later these events are reversed and urinary calcium excretion is raised. Physical inactivity (which sometimes results from the cramped quarters in submarines and spacecraft) exacerbates the calcium loss from bone, as does a reduced vitamin D concentration consequent on lack of sunlight in submariners.

Any toxic gases in the atmosphere must be removed, as well as dust, bacteria and odours. Carbon monoxide and other fumes from machinery can be major hazards if allowed to accumulate, but these are usually removed by a combination of combustion, filtration and absorption.

With prolonged enclosure, such as will occur if man is to explore distant parts of the solar system, recycling of materials will be necessary. Photosynthesis by algae, which utilize carbon dioxide to produce oxygen, may have an important role to play, especially since they may also synthesize nutrients from human waste nitrogenous materials.

Neonates in an incubator have fewer problems but sometimes with ill infants the PO_2 has to be raised. Originally 100 per cent oxygen, or close to it, was used, but this was observed to cause fibrosis behind the lens of the eye (retrolental fibroplasia) with eventual blindness. Therefore, PO_2 is usually kept below 300 mmHg (40 kPa) and even this pressure is maintained for only short periods of time. In incubators, there is a flow-through gas system venting to the outside atmosphere and hence there is no problem with excess carbon dioxide.

8.3 Temperature

Adults feel uncomfortable if they sweat profusely or shiver violently and so it is advantageous to maintain their immediate surroundings (generally the air trapped in clothing) within the thermoneutral zone (see 5.1.1). 'Comfort', however, depends on an interaction between deep body and skin temperatures and normally adults feel comfortable with a skin temperature of about 33°C. If deep body temperature is below the 'set point' (i.e. the 'ideal' value as determined by the hypothalamus), then the skin temperature of 33°C will be perceived as too cold and regulatory mechanisms (such as shivering) are set in motion to raise body temperature. Thus the anomalous situation may exist in which a patient has an increased body temperature but is shivering and thinks that he or she is cold. If the deep body temperature is above the 'set point' (as occurs during resolution of a fever), then sweating occurs and a cool skin is preferred. Subjective comfort, as well as body temperature, shows a circadian rhythm (see 6.1 and 6.3.1). In the evening, body temperature is falling as a result of a decrease in the 'set point' temperature. Deep body temperature falls slowly compared with the decrease in 'set point' because of the thermal capacity of the body; at the same time skin temperature rises (since heat loss occurs by vasodilatation). The subject therefore feels warm and prefers a cool environment. Just after rising, the converse is true and the subject prefers a warmer environment.

Clearly, for adults, the most comfortable temperature of the artificial environment is difficult to determine.

For neonates the problem is rather different. Most babies, especially premature ones, are placed in incubators because they have difficulty in maintaining body temperature. This arises partly because the thermoneutral zone is higher in neonates (see 2.5.1). Premature babies, particularly ill ones, are usually nursed naked, or near naked, and the problem of heat loss can only be overcome by raising the air temperature within the incubator. The design of incubators is such that air flows over the baby and sufficient movement occurs to mix the air and maintain a stable temperature. This air movement will result in a large loss of water from the exposed skin and this can compromise an already precarious fluid balance. Humidity as well as temperature must therefore be controlled. Again the precise temperature at which the incubator should be set is difficult to determine and the baby's body temperature needs to be monitored for optimal control. Care must be taken to avoid excessive radiant heat which, through the perspex canopy, could greatly increase the temperature inside the incubator.

8.4 The daily routine

The enclosed environment may produce psychological stress. First, individuals may be working at abnormal hours and in uncomfortable and dangerous conditions. Secondly, they are separated from their familiar surroundings and loved ones. It has even been suggested that placing neonates in an incubator causes them stress. In these isolated surroundings some adults spend much of their time brooding on the hazards of the mission and their heavy reliance on machinery and technology; some become socially withdrawn; some contemplate spiritual matters. All have to endure partial loss of privacy and the constant presence of others. Relationships can become strained and tempers lost. Clearly, means of maintaining morale and decreasing stress are necessary. Some of these are now outlined.

8.4.1 *Work–rest schedule*
Where possible, the work–rest schedule should remain in phase with the home environment. This helps to preserve contact with 'home' but also stabilises circadian rhythms. For the same reasons, if shift work is being performed, regular schedules are preferable to those that are irregular since there is some evidence that stable circadian rhythms reduce the malaise associated with shift work.

8.4.2 *Diet and personal hygiene*
A palatable and varied diet is important and, where possible, individual preferences should be taken into account. If recycled materials are to be

used (see 8.2), the food must be made acceptable and occasionally its origins disguised. Space and facilities, however, are frequently limited. Much food has to be reconstituted from dehydrated stock and this poses a greater problem in outer space than in submarines. Imaginative catering is important in any prolonged stay in an abnormal environment.

In an enclosed environment, personal cleanliness is highly desirable but motivation may be lacking. In the weightlessness of space flights there is the added inconvenience that washing, micturition and defaecation all require special apparatus.

8.4.3 *Recreation and exercise*

After duty in a restricted environment, personnel may feel that they want to 'get out into the fresh air'. In addition to this psychological need there are physiological reasons — especially marked in weightless conditions — why prolonged inactivity is undesirable (see 7.3.4, 7.3.5). The provision of exercise facilities has therefore become routine and the requirement to use these has been built into work–rest schedules during space flight. However, some astronauts have complained of the tedium and stress associated with lengthy exercise schedules. Similar complaints are voiced by submariners; but exercise is essential to maintain fitness (especially in fighting men), to prevent disuse atrophy of the muscles and to minimize calcium loss from the bones. An added problem is that any exercise will increase the load on the systems controlling the oxygen and carbon dioxide levels in the artificial atmosphere.

Leisure time can be spent learning or appreciating literature and music. Especially among older personnel on long submarine voyages, there is a tendency to use spare time creatively. Thus, an individual's aspirations — so often unrealized in normal life with all its distractions — can be fulfilled and a constructive use for such an abnormal environment can be found.

8.5 **Further reading**

Miles, S. & Mackay, D.E. (1976). *Underwater Medicine Environments*, 4th edn. Adlard Coles: London.

Chapter 9

Exercise

Summary

Exercise can be classified as static ('isometric') or dynamic, and as brief or sustained: each makes different physiological demands.

In sustained dynamic exercise, energy is generated by oxidation of fat, supplemented by oxidation of carbohydrate. At most, only 25 per cent of the original chemical energy is realised as useful mechanical work. Intensity of sustained exercise is limited by the capacity of the muscle for oxidative phosphorylation, and by the capacity of the cardiorespiratory system to supply oxygen. Oxygen uptake in muscle is determined by the amount extracted from circulating blood and by the total blood flow. Extraction is increased by a low tissue PO_2, a shift to the right of the oxyhaemoglobin dissociation curve, and redistribution of capillary blood flow.

Arteriolar dilatation in muscle and constriction in other vascular beds divert more of the cardiac output to muscle, and improved venous return permits a greater cardiac output via a faster heart rate. Pulmonary oxygen uptake increases because the lungs receive a larger flow of more desaturated blood from the muscles; ventilation increases proportionately.

At the beginning of exercise, and during very heavy exercise, energy consumption exceeds pulmonary uptake of oxygen. This 'oxygen deficit' is covered by depletion of stores of oxygen and creatine phosphate, and by anaerobic glycolysis. Only a limited quantity of energy can be supplied and is repaid after exercise as an 'oxygen debt'.

The cardiorespiratory changes that occur in exercise cannot be wholly explained by baroreceptor and chemoreceptor reflexes. There appear to be other central and peripheral stimuli associated with movement which reset the level of ventilation and autonomic activity. These vary with the type and severity of exercise.

Hyperthermia and fluid loss may terminate exercise in hot environments, and lactic acidosis limits very severe exercise.

Some effects of training are specific to an activity or muscle group (such as greater muscle strength, or mitochondrial and capillary density). Others (increased stroke volume and cardiac output) occur in most types of endurance training. Clinically, safe levels of exercise should be judged by the subject's response rather than by reference to normal standards.

Daily energy expenditure varies widely, both between occupations and within a day. Peaks of strenuous exertion are interspersed with periods of rest or light activity, thereby reducing the average rate of work. A similar pattern applies to most sports.

9.1 Introduction

9.1.1 *Exercise*

Exercise is a part of everyday life and needs the same kind (if not the same scale) of physiological adjustments as running a marathon. Formal exercise on a squash court, a running track or a laboratory treadmill is just an extension of a normal activity. To the physiologist, it is a natural way of applying stress to the body's homeostatic mechanisms. Although something can be learned about such mechanisms by observing them in a state of equilibrium, far more can be learned by deliberately disturbing that state and watching them readjust to a new equilibrium.

Nobody has an infinite capacity for exercise. What sets the limit? This is not normally an easy question to answer, since the physiological processes that co-operate to supply energy are well matched. However, in a patient with a cardiorespiratory disorder (who may show little distress at rest), even trivial exercise may expose a weak link quite dramatically. Thus exercise can be useful clinically, in arriving at a diagnosis and in assessing both severity of disability and subsequent progress.

In this chapter, given values generally apply to young adults who are healthy, but not highly athletic, and probably male. The main quantitative differences in females appear to follow from, on average, a smaller total muscle mass and a lower haemoglobin concentration in blood.

9.1.2 *Types of exercise*

It is useful to classify exercise in terms of the physiological adjustments it requires. For example, static ('isometric') exercise such as weight-lifting is quite different from dynamic exercise such as rowing. Though similar muscle groups may be active in both, the patterns of contraction, or of energy consumption and of blood flow, are not the same.

Prolonged static exertion is uncommon (unless maintenance of posture is included), but both short spurts and sustained flows of energy are used in most forms of physical activity. Biochemically, the former are generated largely by anaerobic processes. Since the time-span is brief, there is little scope for cardiorespiratory adjustments during the exercise itself, though there may be anticipatory changes and there will be retrospective changes. A sustained flow of energy lasting 10 min or more can be generated only by aerobic processes. These, in turn, rely directly upon cardiorespiratory adaptation to the increased demand for oxygen. Of course, the two forms of exercise and energy supply are not mutually exclusive. It is easier to deal first with the aerobic and adaptive processes that can sustain 'steady-state' exercise, and later with the anaerobic, anticipatory and retrospective aspects. When exercise continues for longer periods (from half an hour to several hours), disturbances of body temperature and fluid balance and depletion of substrate for energy production may also limit performance.

9.2 Aerobic and adaptive aspects

9.2.1 *Sources of energy*
9.2.1.1 *Adenosine triphosphate (ATP) and its regeneration*. When the cross-bridges are formed between actin and myosin in the course of muscular contraction, the high-energy compound ATP is hydrolysed to adenosine diphosphate (ADP) and inorganic phosphate (Pi). Under aerobic conditions, ATP can be resynthesized from ADP and Pi by oxidative phosphorylation. Alternatively, ADP can be rephosphorylated to ATP by transfer of a phosphate group from other high-energy compounds such as creatine phosphate or by substrate-level phosphorylation during glycolysis. Transfer from creatine phosphate is just an 'internal loan' and there is no net gain of high-energy compounds; its role is discussed later under anaerobic aspects (see 9.3.1.2). Substrate-level phosphorylation does result in a net gain, and since it does not rely directly upon the presence of oxygen, can also take place anaerobically.

9.2.1.2 *Pathways of energy production*. Both carbohydrate and fat are used as substrates for energy production in skeletal and cardiac muscle. The relative amounts used vary with availability of the substrates, type of muscle, muscular activity and supply of oxygen. In resting muscle and in muscles which are active over long periods (such as 'red' skeletal and cardiac muscles), fat is the major substrate under aerobic conditions.

The pathways involved in ATP synthesis are summarized in Figure 9.1. Under aerobic conditions, the overall yield is 16 ATP molecules per 2-carbon unit of a fatty acid and 38 ATP per molecule of glucose, or 39 ATP per molecule of glucose-6-phosphate derived from glycogen.

Under anaerobic conditions, glycolysis halts at pyruvate. However, the accumulating pyruvate can be reduced to lactate (much of which enters the blood) thereby oxidizing $NADH.H^+$ to NAD^+. Though no ATP results from this reaction, it does permit glycolysis to continue, yielding a net 3 ATP per molecule of glucose-6-phosphate. The usefulness of this grossly uneconomic reaction is discussed later under anaerobic aspects (see 9.3.1.3).

9.2.1.3 *Control of energy production*. Given sufficient oxygen and substrate, the rate of oxidative phosphorylation is determined by the supply

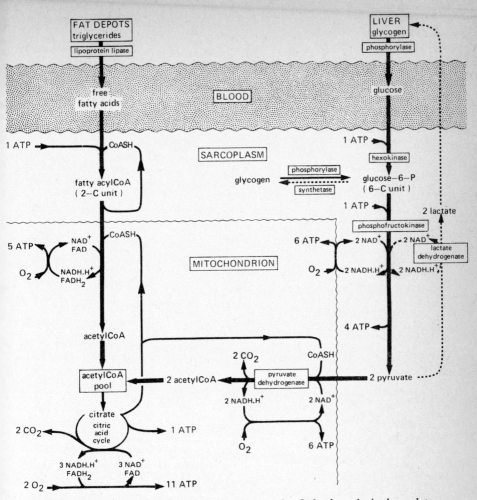

Fig. 9.1 Pathways of energy production in muscle. Only the principal regulatory (rate-limiting) enzymes are included. Alternative pathways are shown in broken lines.

of ADP and Pi, which in turn reflects the rate at which ATP is used. The only limitation is the mitochondrial capacity for oxidative phosphorylation. This capacity is greatest in cardiac muscle and least in 'white' skeletal muscles which, typically, are active in short bursts and maximal efforts. The latter have a higher glycogenolytic capacity.

In general, the extent to which a muscle uses fat parallels its capacity for oxidative phosphorylation and hence, its mitochondrial density. Energy production from carbohydrate can be regarded as an 'optional extra', used to supplement the supply from fat as required. Its advantages are that the large intramuscular stores of glycogen can be used as substrate at short notice, and that a limited amount of energy can be supplied anaerobically. Its disadvantages are that it uses a scarce resource and that anaerobic glycolysis is wasteful of energy, even when lactate is recycled.

The principal internal controls act through product inhibition of key enzymes. For example, phosphofructokinase is inhibited by citrate and pyruvate dehydrogenase by acetyl-coenzyme A; both are inhibited by ATP. Thus, when oxidation of fat is already generating sufficient acetyl-CoA and ATP, glycolysis is suppressed. Phosphorylase is inhibited by glucose-6-phosphate and by ATP, so that glycogenolysis is suppressed when glucose and ATP supplies are adequate.

In exercise, adrenaline may be released. This activates lipoprotein lipase (which mobilizes free fatty acids from fat depots) and phosphorylase in both liver and muscle (which promotes glycogenolysis). In long-continued exercise, glycogen reserves become depleted, and hypoglycaemia occurs. The ability of muscles to use fat is then vitally important.

9.2.1.4 *Efficiency of exercise.* Of the energy released when carbohydrate or fat is oxidized to carbon dioxide and water, the greater part appears as heat. Generation of ATP by oxidative phosphorylation has an efficiency of about 50 per cent, and anaerobic glycolysis is less efficient (even when the lactate is recycled; see 9.3.1.3). Further losses of energy occur as heat when ATP is used in muscle contraction. Overall, only about 25 per cent of the original chemical energy of carbohydrates or fat can be realized as useful mechanical work.

In exercise on a bicycle ergometer, it is relatively easy to estimate the total work done, but in running on the flat or in swimming, most of the work lies in acceleration and deceleration of the limbs and in overcoming viscosity. The apparent efficiency of exercise depends very much on the allowances made for 'invisible' work and for basal energy consumption. Marginal (or 'net') efficiency calculated from the *difference* in energy expenditure between running on the flat and running uphill is obviously higher than the gross efficiency calculated from total energy expenditure. By definition, the efficiency of static (isometric) exercise is nil, since the load is not moved. The energy cost of 'negative' work (where an active muscle is stretched by a load) is far less than in 'positive' work. Mathematically inclined readers may calculate whether the resulting efficiency is large, or negative (or both!).

9.2.2 *Uptake of oxygen by muscle*
The oxygen consumption of muscle, like that of any other tissue, is determined by the Fick principle, namely:

$$\text{Rate of oxygen consumption } (\dot{V}O_2) =$$
$$\text{Blood flow } (\dot{Q}) \times \text{arteriovenous difference of } O_2 \text{ content } (\Delta avO_2)$$

An increase in blood flow alone is an uneconomic and largely ineffective way of increasing oxygen supply. Unless the muscle extracts more oxygen from incoming arterial blood, the main result of a greater blood flow will be a rise in the oxygen content of venous blood. The solution is a typical example of autoregulation: the metabolic changes that occur in exercising muscle induce both increased extraction of oxygen and increased blood flow. The principles governing this short-term regulation of oxygen delivery are essentially the same as those that underlie the long-term adaptation seen at high altitude (see 4.3.5).

9.2.2.1 *Extraction of oxygen from blood*.

Transfer of oxygen from blood in muscle capillaries to mitochondria within muscle cells occurs by diffusion. The steeper the gradient of partial pressure of oxygen from blood to tissue, the faster the diffusion. The main variables affecting this gradient are the partial pressure of oxygen (PO_2) in capillary blood, the PO_2 in the mitochondria and the distance oxygen must diffuse.

At rest the PO_2 in mitochondria is close to that in the venous blood from the muscle, i.e. 20–30 mmHg (3–4 kPa). However, during exercise it may fall to 1 mmHg (0.15 kPa) before mitochondrial activity is impaired. This in itself makes the gradient of PO_2 steeper and speeds diffusion of oxygen from blood; that is, the demand accelerates the supply. The presence of myoglobin, in greater concentration in 'red' muscles, also assists diffusion within muscle.

Haemoglobin in venous blood from resting muscle is 40–60 per cent saturated but in maximal exercise saturation falls to 15 per cent or less. This greater desaturation reflects a lower venous PO_2 of about 15 mmHg (2 kPa) and also a displacement of the oxyhaemoglobin dissociation curve to the right, by products of metabolism such as CO_2, H^+ and heat (see 4.3.4, Figure 4.6). In femoral venous blood of a trained subject, P_{50} at rest was 26 mmHg (3.5 kPa) but rose to 39 mmHg (5.2 kPa) during maximal leg exercise. PO_2 fell to 10 mmHg (1.3 kPa) and saturation to 5 per cent. Thus, in maximal exercise, ΔavO_2 can be twice that at rest.

It follows that oxygen is diffusing from blood to muscle at least twice as fast, despite a lower capillary PO_2. In part, this results from the fall in the PO_2 of the tissues (above), but the third component of the diffusion gradient (the distance) also changes. This depends on the distribution of blood flow in the capillary network.

9.2.2.2 *Blood flow in muscle*.

In resting muscle, tonic sympathetic nervous activity constricts arterioles and reduces blood flow to 20–30 $ml \cdot min^{-1} \cdot kg^{-1}$. The distribution of this flow within the microcirculation is determined by the opening and closing of pre-capillary sphincters in response to local metabolism. At rest the majority of sphincters are closed, and most of the flow is confined to main or 'thoroughfare' channels, so that effective capillary density is probably about 100 mm^{-2}.

In working muscles, PO_2 falls, whereas the partial pressure of carbon dioxide (PCO_2), temperature, osmotic activity and the concentrations of H^+, K^+ and ADP in interstitial fluid all rise. These local factors relax pre-capillary sphincters and override the centrally controlled vasoconstriction of the arterioles. Some selective sympathetic vasodilatation may also occur (see 9.3.2). As a result of the reduced resistance, total blood flow rises towards a maximum of about 500 ml·min^{-1}·kg^{-1}. Relaxation of the sphincters and the rise of perfusion pressure open dormant capillaries: estimates of capillary density in human quadriceps femoris during maximal exercise are in the range 300–700 mm^{-2}. The greater total flow is therefore spread more evenly across a larger number of capillaries (Figure 9.2). Since the flow has increased 20-fold and the number of capillaries about fivefold, the flow rate through individual capillaries must be more rapid than at rest. On the other hand, the distance the oxygen must diffuse is now shorter; diffusion is more rapid and the blood is efficiently desaturated before it leaves the capillary. Overall, then, total oxygen uptake by muscle has increased 40-fold. There are also changes in fluid balance across the capillary wall (see 9.4.1.2).

A notable exception to this pattern occurs in static ('isometric') exercise. As the force of a sustained contraction increases, the intramuscular vessels are compressed and blood flow decreases. At maximal tension, the muscle is totally ischaemic and dependent on anaerobic energy. In contrast, tensions of 10–15 per cent of maximum can be maintained for long periods; this is presumably the situation in postural muscles.

9.2.2.3 *Implications for peripheral circulation.* Skeletal muscle constitutes 40–45 per cent of lean body mass. At rest, it receives less than 20 per cent of the cardiac output but, if maximally active, it would require 300–400 per cent. Clearly, if no compensatory action were taken to meet the increased demand for blood flow, arterial blood pressure would fall. In practice, an initial fall of blood pressure does not occur because anticipatory mechanisms forestall it (see 9.3.2). For the present, we will ignore these and suppose that a fall has occurred and that corrections have been initiated through the baroreceptor reflex pathway.

An obvious response is an increase in cardiac output to match the increased demand. However, this cannot be achieved instantaneously because it requires an increased venous return (see 9.2.3.1). More imme-diately useful is a generalized vasoconstriction resulting from increased sympathetic activity. Some tissues are spared, namely the high-priority

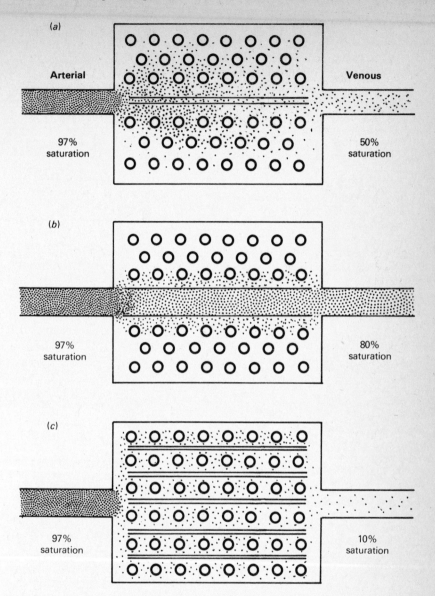

Fig. 9.2 Blood flow in capillaries of muscle. Tissue PO_2 declines with radial and longitudinal distance from the arterial (left) end of the capillary. In resting muscle (a), PO_2 at remote cells is low but adequate. In working muscle, oxygen uptake by cells close to the capillary increases and PO_2 declines more rapidly with distance; remote cells are hypoxic. In (b) faster blood flow shortens the time available for extraction of oxygen and extraction is less complete. In (c) the increased flow is redistributed across a larger number of capillaries: each cell is closer to a capillary, velocity of flow in individual capillaries is slow and extraction of oxygen is high.

circulations of brain and heart which are permanently protected against outside interference, and the working muscles where local metabolic factors predominate. Other tissues have no such protection: the principal victims are the splanchnic (gut and liver) and renal circulations (where blood flow in maximal exercise can be as low as 30 per cent of normal), the skin and non-working muscles. The saving is up to half the resting cardiac output, or enough to increase total muscle blood flow threefold. This may seem a small gain, which could not be maintained without damage to the vasoconstricted organs. Even so, it is useful in bridging the gap before cardiac output rises, and in ensuring that none of the increase is wasted upon these (temporarily) irrelevant organs. Blood flow to skin may later rise again if body temperature increases (see 9.4.1.1).

9.2.3 *Cardiac output*

The heart meets the demands of the tissues for blood flow; it does not impose flow upon them. This service is provided by adjusting cardiac output to maintain a sufficient perfusion pressure irrespective of the total peripheral resistance. A small or temporary increase in demand by one tissue may be compensated by constriction elsewhere, so that the total resistance is unaltered. A large, continued increase in demand necessitates an increase in cardiac output. However, the heart cannot sustain a greater output until it receives a greater input. A rise in heart rate, alone, may increase left heart output for a few beats by drawing upon blood in the pulmonary veins, but the effect will not last unless there is an increase in systemic venous return.

9.2.3.1 *Venous return.*

Cardiac output relies upon recirculation of a more-or-less fixed volume of blood. Venous pooling must be prevented: the less time blood spends travelling back through the veins, the more often it can be recirculated. In exercising limbs, this is usually automatic, via the 'muscle-pump' effect. When limb muscles contract rhythmically, the deep veins are intermittently compressed and, provided their valves are competent, the blood is driven towards the heart. Once blood reaches the central veins, other mechanisms take over. One is the pumping effect produced by changes in differential pressure between abdomen and thorax during breathing. In inspiration, thoracic pressure falls, whereas abdominal pressure rises, compressing the abdominal veins and driving blood into the thorax. This effect is amplified by the greater frequency and depth of breathing in exercise. The other major mechanisms are vasomotor. The splanchnic and renal circulations, at rest, account for 50–55 per cent of cardiac output and hold about 25 per cent of blood volume. We have already seen that, in maximal exercise, much of the blood flow can be diverted to muscle. In addition, there is a marked generalised venoconstriction, which reduces the capacitance of the venous system as a whole, and that of the abdominal veins in particular.

9.2.3.2 *Heart rate.* In a normal subject, heart rate rises approximately linearly with the severity of exercise, towards a maximum of about 200 beats·min^{-1} in a young adult (Figure 9.3). (Maximum heart rate declines with age to about 160 beats·min^{-1} at 60 years.) The progressive rise is due at first to decreasing vagal activity and then to a growing sympathetic drive.

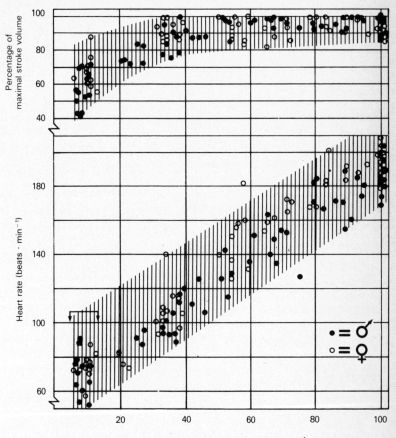

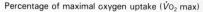
Percentage of maximal oxygen uptake ($\dot{V}O_2$ max)

Fig. 9.3 Stroke volume and heart rate in exercise. Data from 23 subjects working in the sitting position on a bicycle ergometer. Points between the arrows represent observations at rest. Note that most of the increase in stroke volume occurs from rest to 40 per cent of maximal oxygen consumption ($\dot{V}O_2$ max), whereas heart rate increases linearly with work load. Redrawn from Åstrand, P.-O., Cuddy, T.E., Saltin, B. & Stenberg, J. (1964). *Journal of Applied Physiology* **19**, 268–74.

The latter is probably mainly nervous, though circulating catecholamines can produce a substantial increase in rate of a denervated heart in exercise. The causes of the change in autonomic activity are less clear. Like the generalized vasoconstriction and venoconstriction described earlier, it could be initiated by baroreceptor reflexes if vasodilatation in working muscles led to a fall of arterial blood pressure. However, the rise in heart rate closely parallels the peripheral vasoconstriction, with respect both to time and to severity of exercise. Presumably both effects are due to the same antici- patory mechanisms (see 9.3.2). As the heart rate increases, diastole shortens more than systole (thereby reducing the time available for ventricular filling, and for coronary blood flow). Conduction within the heart is faster (shown by a shorter P–R interval in an electrocardiogram) and there is some shortening of systole through an increased velocity of contraction of cardiac muscle.

9.2.3.3 *Stroke volume*. In many cases, there must also be an increase in stroke volume, since cardiac output often rises more than would be expected from the increase in heart rate alone. The changes in stroke volume, though, follow a different pattern from those in heart rate (Figure 9.3). The most marked rise occurs when a subject in the upright position undertakes moderate exercise. There is only a small further rise in maximal exercise, to about one-and-a-half times the resting value. In the supine position, the stroke volume at rest is already larger than when upright, and increases less when the subject exercises.

The mechanism of the increase is a source of much debate. A traditional explanation is based on Starling's 'Law of the Heart': greater filling by a higher venous pressure leads to a more forcible systole. However, there is no consistent relationship between right atrial pressure and stroke volume, and radiographic studies do not show any significant increase in end- diastolic volume in exercise. A more likely explanation is that there is a sympathetically produced increase in contractility, leading to more complete emptying in systole. Other factors may be the reduced peripheral resistance, and more efficient filling by a faster venous return. The non- linear increase in stroke volume, and the effects of posture, suggest that it is not a closely controlled single effect but rather the resultant of several factors.

9.2.3.4 *Arterial blood pressure*. Systolic pressure can rise to over 225 mmHg (30 kPa) in some kinds of exercise, though 190 mmHg (25 kPa) is a more typical limit. Diastolic pressure, on the other hand, shows only a small rise and in some subjects may fall slightly. Thus pulse pressure may rise two- to threefold (Figure 9.4). The high systolic pressure results from ejection of the same (or larger) stroke volume in a shorter time, whereas the

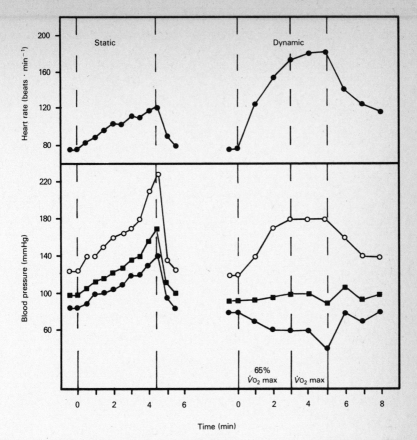

Fig. 9.4 Heart rate and arterial blood pressure (○—○, systolic; ●—●, diastolic; ■—■, mean) in static (isometric) and dynamic exercise. In static exercise, the subject maintained a hand-grip at 30 per cent of maximal voluntary contraction (MVC); in dynamic, he walked on a treadmill at submaximal and maximal rates of oxygen consumption ($\dot{V}O_2$ max). Note that, in dynamic exercise, the mean blood pressure rose little despite a large rise in heart rate; by implication, there must have been a large fall in peripheral resistance. Originally published in Lind, A.R. & McNicol, G.W. (1967) *Canadian Medical Association Journal* **96**, 706–15.

diastolic pressure reflects the close balance between increased cardiac output and reduced peripheral resistance. In static ('isometric') exercise, or in maximal exercise of a small mass of muscle (e.g. arm exercise), the rise in diastolic pressure is much greater. This is probably a consequence of an excessive sympathetic activity (see 9.3.2.4).

9.2.3.5 *Coronary blood flow.*
The work of the heart increases in exercise approximately in proportion to cardiac output in dynamic exercise (but

disproportionately so in static exercise). Its oxygen consumption increases accordingly, mainly through increased coronary flow rather than by greater extraction from its already highly desaturated venous blood; this is despite the shorter time available for coronary flow (see 9.2.3.2).

9.2.4 *Pulmonary oxygen uptake*

Oxygen uptake in the lungs, like that in the muscles, is the product of blood flow (in this case, cardiac output) and arteriovenous difference of oxygen content (ΔavO_2). The 'arterialised' blood leaving the lungs is normally 95–97 per cent saturated. Until a larger flow and/or more desaturated 'mixed venous' blood reaches the lungs, oxygen uptake cannot increase, whatever the increase in ventilation (except in so far as the thoraco-abdominal pump hastens venous return).

9.2.4.1 *Flow and saturation of mixed venous blood*. At rest, there is a large venous flow from the renal and splanchnic circulations, which extract relatively little oxygen, and a smaller flow from skeletal and heart muscle, of lower oxygen content. The oxygen saturation of the mixture reaching the lungs is therefore high (70–75 per cent (Figure 9.5)). In exercise the splanchnic and renal flows decrease (and are probably more desaturated), whereas muscle flow increases greatly (and is even more desaturated). The mixed venous oxygen saturation falls towards that of exercising muscle and may reach as low as 20–25 per cent. Thus each litre of cardiac output can accept up to three times more oxygen in the lungs than at rest, say 6–7 mmol (150 ml). Given a cardiac output four times that at rest, say 20 litres $\cdot$ min^{-1}, oxygen uptake can rise to 120–140 mmol (3 litres) per minute.

9.2.4.2 *Changes in pulmonary factors*. There are three pulmonary factors that might limit pulmonary gas exchange and lead to an arterial oxygen saturation of less than 95–97 per cent: ventilation, diffusion, and matching of ventilation and perfusion. Ventilation does not appear to impose a limit, since healthy subjects are capable of levels of ventilation higher than the 100–150 litres $\cdot$ min^{-1} of maximal exercise. In disorders that increase the work of breathing (e.g. airways obstruction), the effort required to breathe and the consequent oxygen consumption by the ventilatory muscles can severely limit exercise capacity.

The pulmonary capillary bed is unlike that of resting muscle: there are few if any fully dormant capillaries. Instead, there is a marked regional variation in perfusion and blood volume, capillaries at the apices receiving little blood. When cardiac output increases in exercise, pulmonary blood pressure rises. The rise is small in absolute terms, 7–15 mmHg (1–2 kPa), but this is sufficient to produce a disproportionately large increase in apical

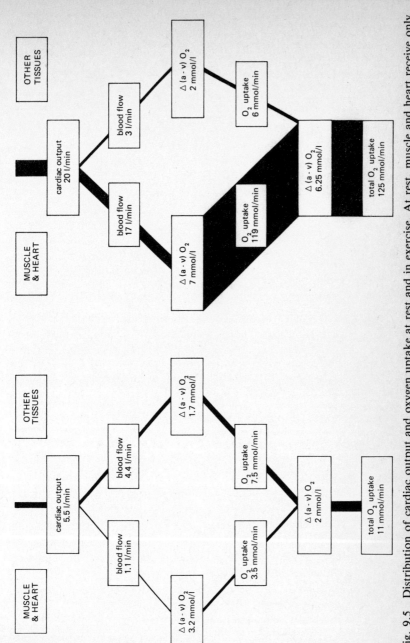

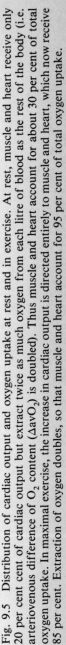

Fig. 9.5 Distribution of cardiac output and oxygen uptake at rest and in exercise. At rest, muscle and heart receive only 20 per cent of cardiac output but extract twice as much oxygen from each litre of blood as the rest of the body (i.e. arteriovenous difference of O_2 content (ΔavO_2) is doubled). Thus muscle and heart account for about 30 per cent of total oxygen uptake. In maximal exercise, the increase in cardiac output is directed entirely to muscle and heart, which now receive 85 per cent. Extraction of oxygen doubles, so that muscle and heart account for 95 per cent of total oxygen uptake.

blood flow and volume. This affects both diffusion and ventilation/perfusion.

The average velocity of flow in individual capillaries approximately doubles, but since at rest blood reaches equilibrium with alveolar air before it is halfway through the capillary, there is still sufficient time for equilibration. The low PO_2 of the incoming blood also speeds diffusion. The larger volume of blood in apical capillaries increases the diffusing capacity about twofold, though, like changes in stroke volume (see 9.2.3.3), this effect depends on posture and is not linearly related to intensity of exercise. In disorders where diffusing capacity is barely adequate at rest (pulmonary fibrosis, emphysema), there is no reserve, and even light exercise results in a fall in arterial saturation.

In health, underperfusion of the apices at rest leads to only a small inefficiency of gas exchange. Probably, we should regard the increased perfusion during exercise as recruitment of 'spare' capacity.

9.2.4.3 *Control of breathing*. Over a wide range of intensity of exercise, ventilation rises in close proportion to oxygen consumption and carbon dioxide production, and arterial PO_2 and PCO_2 show only small, erratic changes from resting levels (about 7 mmHg (1 kPa)). Towards maximal intensity, ventilation rises faster than oxygen consumption and PCO_2 falls.

What causes breathing to increase in exercise? Though chemoreceptor reflexes (stimulated by rising PCO_2 or falling blood pH or PO_2) certainly can drive ventilation, the changes in these stimuli in exercise seem too small and variable to account for the large consistent increase in ventilation. The problem is analogous to that described earlier for baroreceptor reflexes in relation to cardiovascular changes: again 'anticipatory' controls may be responsible (see 9.3.2).

9.3 Anaerobic and anticipatory aspects

We have considered the demands of 'steady-state' exercise, yet this is unusual in everyday life, where the level of activity is continuously changing. The transition from rest to exercise raises two major issues: what happens when the demand for energy exceeds the aerobic supply and how are cardiorespiratory adjustments initiated in the apparent absence of stimuli to baroreceptors and chemoreceptors?

9.3.1 *Anaerobic energy sources*
9.3.1.1 *'Oxygen deficit' and 'oxygen debt'*. There is a long chain of cause-and-effect between breakdown of ATP in working muscle and pulmonary uptake of oxygen; in the meantime, the muscle is running short of ATP. This problem arises at the onset of any exercise, however mild. It is

particularly acute in very severe exercise, which may exceed even the maximal aerobic capacity. Energy must be drawn from other sources

(a) to cover the initial deficit of energy until the supply of oxygen catches up, and

(b) to supplement the aerobic supply where this is inadequate.

Energy 'borrowed' from these other sources must eventually be repaid, often 'with interest'. The energy to pay this total 'debt' (of 'deficit' plus 'interest') is 'earned' in the form of additional oxygen consumption after exercise.

The true 'oxygen debt' is small: the total oxygen content of the body is lower in exercise because the venous blood is more desaturated and some oxygen is extracted from myoglobin. In maximal exercise, such depletion of oxygen stores might supply 15–20 mmol (300–500 ml) of oxygen, equivalent to 7–9 kJ of aerobic energy. Most of the energy, though, is borrowed from two anaerobic sources: high-energy organic phosphates, and anaerobic glycolysis.

9.3.1.2 *High-energy phosphates*. One of these, ATP, directly supplies energy for muscle contraction and is hydrolysed to ADP in the process (see 9.2.1.1). The other, creatine phosphate, is present in a concentration three to four times that of ATP in resting muscle and acts as a reserve from which ATP is regenerated. In heavy exercise, ATP concentration may fall by up to 30 per cent whereas creatine phosphate may fall by 80 per cent, thus supplying energy equivalent to consumption of 60–70 mmol (1.5 litres) of oxygen, or 30 kJ of aerobic energy. This source has the advantage of being instantly available and rapidly recharged after exercise (with a half-time of 20–30 s).

9.3.1.3 *Anaerobic glycolysis*. In Section 9.2.1.2 it was noted that glycolysis can yield ATP in the absence of oxygen. The yield is small compared with aerobic metabolism (2 or 3 ATP per 6-carbon unit of carbohydrate versus 38 or 39 ATP), and the relatively strong acid, lactic acid, accumulates. It plays a larger role in 'white' muscle (which has a high glycogenolytic capacity and is recruited in brief maximal efforts) than in 'red' (which has a higher aerobic capacity and is active over long periods). Though anaerobic glycolysis is most obvious when the demand for energy exceeds the aerobic capacity, blood lactate levels show a rise even when exercise intensity is only 50–60 per cent of maximal aerobic capacity.

In the short term, the process is limited by the body's tolerance for lactic acid and the fall in pH it produces (see 9.4.2.2). Typically, blood lactate concentration rises from a resting value of 1–1.5 mmol $\cdot$ litre^{-1} to a peak of 10–15 mmol $\cdot$ litre^{-1}. When a large mass of muscle is active, 0.7–1.0 mol of lactate can be produced, equivalent to 80–115 kJ of aerobic energy or 170–250 mmol (3.5–5.5 litres) of oxygen consumed.

Lactate still in muscle at the end of exercise may revert to pyruvate and be oxidized through the citric acid cycle. However, the majority escapes into the blood. Some is oxidized by cardiac muscle, but the greater part (85 per cent) is synthesized into glycogen in the liver. This requires much more energy than the anaerobic glycolysis yielded, so that the final cost of repaying the 'debt' in terms of oxygen consumption is about twice the original 'deficit'. Lactate production is slower in onset than depletion of creatine phosphate, taking 10–15 s to occur, and recovery is very slow, taking an hour or more. During recovery, the depleted carbohydrate stores are conserved by a shift towards oxidation of fat.

An example of the use of anaerobic energy sources is shown in Figure 9.6. It is assumed that the subject reached a steady-state after 8–10 min. The estimated oxygen 'deficit' is 117 mmol, but the 'debt' repaid is 244 mmol; this 'interest' reduced the net efficiency from 24.9 per cent to 22.7 per cent. Components of 'interest' can be seen in the persistence of

(1) low pH, due to lactic acid,
(2) low R, due to oxidation of fat, and
(3) high rectal and skin temperature and heart rate.

In everyday life, it is likely that the fast depletion of oxygen and creatine phosphate can cover most brief or minor changes in physical activity and that anaerobic glycolysis is reserved for major efforts. In fact, it is possible to reach very high average rates of work with very little lactate production or subjective exhaustion, by working in 10–15 s bursts separated by rest

---

Fig. 9.6   Acid–base, metabolic, thermal and cardiac changes during exercise and recovery. The subject worked in the sitting position on a bicycle ergometer at a load of 150 W. The broken horizontal lines represent mean pre-exercise values. Note:

(1) During exercise and much of recovery, arterial blood pH was lower than would be expected from end-tidal carbon dioxide concentration ($F_{ET}CO_2$), implying a non-respiratory acidosis.
(2) During exercise and early recovery, the gas exchange ratio (R) was above unity, implying that more carbon dioxide was being eliminated by ventilation than was being produced by tissue respiration.
(3) Of the oxygen debt 35 per cent was repaid within the first 3 min of recovery and 48 per cent (equivalent to the initial deficit) in 11 min; the remaining debt was repaid very slowly.
(4) Skin temperature (T_S) changed after rectal temperature (T_R); the initial fall in T_R may be due to cooler venous blood from the legs; most of the surplus heat was eliminated during recovery, not exercise.
(5) Heart rate rose and fell rapidly at beginning and end of exercise; the rise immediately before exercise was probably due to anticipation, and the slow decline during recovery may reflect skin blood flow (see (4), above).

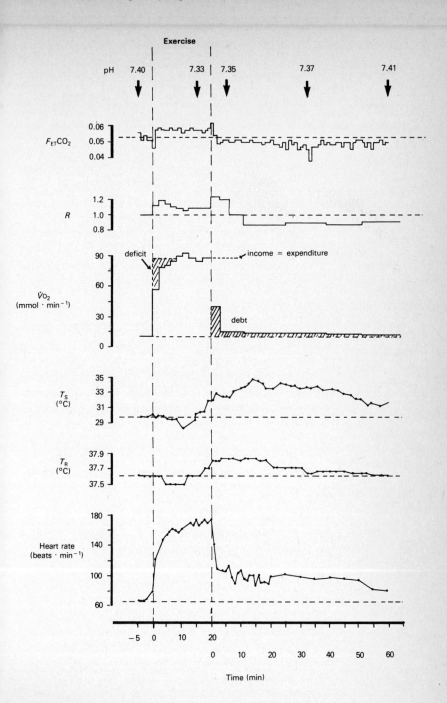

periods of 5–10 s, instead of working continuously at a lower load. Perhaps Aesop's hare was correct in his tactics!

9.3.2 *Anticipatory control mechanisms*

9.3.2.1 *Principle.* As we have seen, some of the adjustments occurring in exercise can be treated as simple cause and effect: increased extraction of oxygen from blood follows almost inevitably from increased oxygen uptake by the muscles. Other adjustments need at least a simple feedback control: stimulation of glycogenolysis by a fall in ATP. A further group could in theory be controlled by simple feedbacks but in practice obviously relies on more sophisticated controls: changes in vasomotor activity, heart rate and breathing. In this last group, adjustments begin before blood pressure or blood gases have changed and reduce (or even reverse) the disturbance before it has happened. There are three ways in which an 'early warning' could occur in exercise: conscious anticipation, central co-ordination, and reflex. These are not mutually exclusive, but complement each other at different stages.

9.3.2.2 *Conscious anticipation.* In formal competitive exercise, there is obviously a strong emotional drive, and heart rate and breathing start to increase before the starting gun is fired. In an experimental situation, this effect can easily be induced by a 'count-down' (Figure 9.6). This is not typical of the casual exercise which forms part of everyday life. Habituation occurs, and there will be only an occasional emotional preface. However, even everyday activities are initiated consciously and are organised at subconscious levels before they take place. Though neurophysiological research in this field has generally been more concerned with co-ordination of the movements than with the cardiorespiratory adjustments needed to support them, there is evidence of similar co-ordinated patterns of cardio-respiratory activity.

9.3.2.3 *Changes at the beginning and end of exercise.* Within the first 5–10 s of starting exercise, heart rate rises about 10–15 beats $\cdot$ min^{-1} (Figure 9.6); this is probably due more to decreased vagal tone than to a sympathetic drive. There is often an even faster but more erratic increase in ventilation of 5–20 litres $\cdot$ min^{-1} (Figure 9.7). What happens next depends very much on the severity of exercise. In very light exercise, there may be no further rise, or even a slight fall in both variables. In moderate exercise, there may be a brief hiatus before slower but larger and more progressive rises towards steady levels. Most of the change occurs in 2–3 min, but there is a further gradual change up to 5–10 min. In heavy exercise, the hiatus is not noticeable. In fact, the initial rapid change in ventilation may not be seen, perhaps because the muscles of breathing are compromised by the

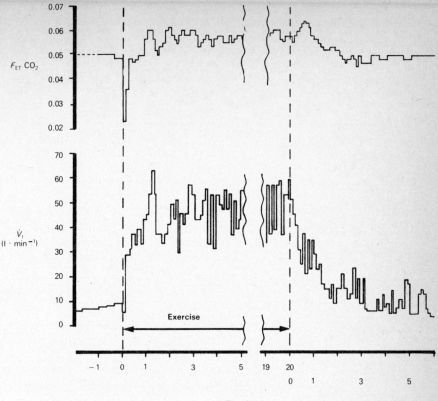

Fig. 9.7 Respiratory changes at the beginning and end of exercise. These are details from the experiment shown in Fig. 9.6. Note:

(1) The brief fall in end-tidal carbon dioxide concentration ($F_{ET}CO_2$) at the beginning of exercise may be an artefact, but ventilation ($\dot{V}_I$) clearly rises before $F_{ET}CO_2$.

(2) The initial peak of $\dot{V}_I$ at 1–1.5 min is not sustained; it is succeeded at 2–3 min by a slower rise which follows the rise in $F_{ET}CO_2$.

(3) $\dot{V}_I$ falls rapidly during the first 0.5 min of recovery, despite the rise in $F_{ET}CO_2$ which results.

mechanical stress of the exercise. At the end of exercise, the changes are more consistent: both heart rate and ventilation fall sharply at first, then more slowly. The rapid changes at the beginning and end are too quick to be explained by changes in the composition or flow of venous blood, and occur even when the circulation to the exercising limbs is occluded. Chemoreceptor and baroreceptor reflexes, if not suppressed, are at least temporarily overridden: at the beginning, ventilation is rising though PCO_2 is unchanged or falling, and heart rate and blood pressure both rise. At the end, ventilation falls though PCO_2 is rising, and heart rate and blood pressure both fall: these relationships are not what would be expected from the usual negative feedback reflexes.

9.3.2.4 *Other evidence*. In static exercise, both systolic and diastolic blood pressure rise progressively to very high values (Figure 9.4). Despite this, heart rate also rises, though not as much as in heavy dynamic exercise: this suggests that baroreceptor reflexes, though overriden, may still partially counteract the imposed tachycardia. The more forcible the static muscular contraction, the larger and faster the rises in blood pressure and heart rate. Occlusion of the local circulation (which may already be impaired by the sustained contraction) tends to enhance the effect.

In dynamic exercise of small muscle groups (e.g. an arm) ventilation, heart rate and blood pressure are higher, for a given rate of oxygen uptake, than where large muscle groups are used. There is also evidence of a 'learning' effect, in that cardiorespiratory changes are smaller after training of the muscles: it is not clear how far this is due to central habituation, or to adaptation in the muscles. Where the central nervous 'command' is dissociated from the muscular force exerted (e.g. by partial curarisation), an increase in central command is accompanied by greater cardiorespiratory changes. Passive and, especially, electrically stimulated limb movements have also been shown to increase breathing, heart rate and blood pressure. Selective blockade, however, suggests that C-fibres (rather than the larger proprioceptive afferents) are principally responsible once exercise is in progress. The stimuli may be chemical changes in the muscles.

9.3.2.5 *Conclusions*. There is probably some central cardiorespiratory activation at the beginning of exercise which is not closely related to the severity of exercise but is linked to initiation of movement. Increased ventilation cannot increase oxygen uptake at this stage, but may pre-empt a lactic acidosis and increase venous return. It is later reinforced or attenuated according to the central 'command' the exercise demands, and also perhaps by information from chemosensitive nerves in muscle. Since, in dynamic exercise, blood pressure appears quite well controlled, though at a higher level, it is likely that baroreceptor reflexes persist but are superimposed on a higher basal sympathetic activity. The role of the respiratory chemoreceptors may be even simpler, namely that they continue to function as at rest. Since they lie in the arterial circulation, they do not detect the load of carbon dioxide reaching the lungs or the level of ventilation, but rather any failure of ventilation to eliminate carbon dioxide. Given that breathing has been elevated by other factors to an approximately adequate level, only minor corrections will be needed to reach a new steady-state.

The control of breathing in exercise is a source of much debate. There are many interpretations more elaborate than the one above, each with some experimental evidence in its favour.

9.4 Other aspects

9.4.1 *Temperature, fluid and acid–base disturbances*
9.4.1.1 *Temperature regulation.* Clearly, the extra heat produced by exercise (see 9.2.1.4) must be disposed of. In a short sprint this is no problem, because the quantity of heat is small and some of it is associated with the repayment of oxygen debt during recovery. At the beginning of exercise, there may even be a small fall of core temperature as venous return from the (initially) cooler limbs increases. This may lead to a fall in skin temperature.

In moderate exercise of longer duration, even in hot environments (e.g. 35°C), the rise in core temperature is only about 1°C. Most of the extra heat is generated outside the body core and a substantial proportion may be eliminated directly from the exercising limbs. Also, sweating begins at a lower core temperature during exercise than at rest. Of course, skin blood flow increases very greatly at the higher environmental temperatures, but cardiac output rises to supply the flow. There is also an increase in heat loss by evaporation from the respiratory tract.

In sustained heavy exercise in a hot environment, the needs of exercise and of thermoregulation conflict. Cardiac output is already maximal and skin blood flow can rise only at the cost of some other tissue. Visceral vasoconstriction intensifies to the verge of hepatic failure through ischaemia and rising temperature. Skin blood flow rises, but not enough, so that core temperature rises progressively. Stroke volume decreases as a result of pooling of blood in cutaneous veins and a reduction in plasma volume (see below). Eventually, cardiac output, arterial blood pressure and oxygen uptake fall. Sometimes, this effect is seen soon after exercise stops: cutaneous blood flow is rising but leg movements have ceased and venous pooling occurs.

9.4.1.2 *Plasma volume; sweating.* In heavy exercise, as much as 15 per cent of plasma volume shifts into the interstitial space of muscle. This is due to increased hydrostatic pressure within the vasodilated capillaries and to accumulation of osmotically active substances in the interstitium. Provided that cardiac output is not affected, this haemoconcentration may somewhat increase delivery of oxygen to the working muscles.

More serious, in exercise of long duration, is the outright loss of water and sodium ions in sweat. In moderate exercise at an ambient temperature of 15°C, the sweat rate is about $0.2–0.3$ litre $\cdot$ h^{-1}, but at 35°C can reach 1.5 litres $\cdot$ h^{-1}. Loss of muscle K$^+$ to the interstitium may also cause weakness.

9.4.1.3 *Acid–base balance.* Carbonic acid is potentially a major threat to acid–base balance since in heavy exercise carbon dioxide production can reach 100 mmol·min^{-1}. Retention of one minute's production would approximately double the hydrogen ion concentration in blood (pH would drop 0.3 units). Normally, of course, carbon dioxide is eliminated as fast as it is produced (or even faster at the beginning of exercise if there is an anticipatory increase of ventilation). Buffering of this larger quantity in venous blood is automatic under aerobic conditions, since haemoglobin is more desaturated.

In practice, the major threat arises from lactic acid produced in the course of anaerobic glycolysis (see 9.3.1.3). When a subject exercises to exhaustion in 1–3 min and employs a large mass of muscle, 0.7–1.0 mol of lactate is formed. During such a short period, there is little time for lactate to leave the muscle, and intramuscular pH falls from a resting value of about 7.0 to 6.4–6.6 at exhaustion. In muscle venous blood, pH may fall to 7.0 or below: this is in part a reflection of the high PCO_2, due both to aerobic production and to buffering of lactate by bicarbonate. In arterial blood (where PCO_2 is lower), the pH reaches a minimum of 7.1 to 7.2 some 5–10 min after the end of exercise. In exercise to exhaustion over a longer time-course, arterial pH may fall even further, presumably because there is more time for lactate to escape from the muscles (though some may then be disposed of metabolically).

The fall in arterial pH stimulates breathing, so that total ventilation increases out of proportion to oxygen consumption and carbon dioxide production by the tissues. The gas exchange ratio (R) rises far above unity as carbon dioxide elimination increases and arterial PCO_2 falls below the resting value. This respiratory compensation is, of course, a temporary expedient: the acid–base disturbance will be fully corrected only when the lactate has been removed during the lengthy recovery from exercise (Figure 9.6). As the lactate is removed, carbon dioxide is retained; this, together with a shift towards oxidation of fat, reduces R below its pre-exercise level.

9.4.2 *Endurance and training*
9.4.2.1 *Endurance.* As one might expect, there is an inverse relationship between intensity and duration of exercise. Maximal efforts, whether static or brief dynamic, are essentially anaerobic. The total force exerted will depend on the muscle mass and its composition: weight lifters and sprinters, for example, have a higher than average proportion of 'white' fibres in their leg muscles. Such an effort can be maintained only for a few seconds — perhaps barely long enough for anaerobic glycolysis to occur. A half-maximal isometric contraction can be sustained for 1–2 min, and a 15 per cent contraction almost indefinitely (see 9.2.2.2).

A similar pattern applies to dynamic exercise. There is a limited quantity of anaerobic energy available (100–150 kJ, see 9.3.1) which will last about 20–30 s at full power: a sprinter runs as fast for 200 m as for 100 m, but is slower over 400 m. Alternatively, the same quantity of energy can be spread over a longer period to supplement, rather than to replace, the aerobic supply. Over 1000 m (2.5 min), roughly half the energy used is anaerobic; the remainder comes from oxidative phosphorylation, which by now has reached its maximum rate. Over 20 km (1 h), the anaerobic energy is spread so thinly that its contribution is negligible. The runner is reduced to an aerobic plateau which is only about 50 per cent of the maximal aerobic capacity in an untrained subject (see 9.3.1.3); a highly trained athletic subject might achieve 80 per cent.

Over longer periods of several hours, there is a further slow decline in power. One factor in this is depletion of glycogen stores: those in muscle reach very low levels in an hour's heavy exercise, and the falling blood glucose concentration suggests that liver glycogen is depleted too. The falling respiratory quotient and rising oxygen consumption confirm the growing dependence on fat. Diet has a marked effect on endurance over these longer periods. Especially if glycogen stores have been previously emptied by exhaustion, a carbohydrate-rich diet can raise muscle glycogen to twice the normal concentration and can correspondingly double the endurance for sustained heavy work.

Disturbances of fluid and electrolyte balance, and the need for thermo-regulation, will also of course limit endurance, particularly in adverse environments (see 9.4.1.1 and 9.4.1.2).

9.4.2.2 *Local changes*. To some extent, the effects of training are specific because a skilled movement is perfected. Irrelevant movement is decreased and energy is expended to better effect. Especially in sustained exercise, there is less recruitment of 'white' muscle and greater reliance on the more enduring 'red' muscles. Even when the skill has been learned, though, major changes continue to occur in the muscles.

An obvious change is that of muscle mass, as in a 'professional strong-man'. This is misleading: such gross hypertrophy is a specialised adaptation to static exercise, though it may also be seen, to a lesser extent, in the leg muscles of a sprinter. The larger muscles can exert larger forces, but not for long. Without corresponding cardiovascular adaptation, dynamic endurance is not increased. It may even be reduced because blood flow is inadequate for the greater mass of tissue. Of course, this hardly matters during a maximal isometric contraction, since the blood flow ceases (see 9.2.2.2). To be effective, training for anaerobic power must place demands on anaerobic energy sources. The subject should exercise to exhaustion as rapidly as possible (e.g. in 1 min) as often as possible (e.g. every 5 min); this is unpleasant.

At the other extreme, middle- and long-distance runners show relatively modest changes in size and force of the leg muscles, but posses very great endurance. In addition to some increase in size of the fibres, a number of biochemical and local cardiovascular changes occur in dynamically trained muscles. In general, these changes result in a greater capacity to produce energy aerobically.

The biochemical pattern varies with the type of muscle fibre. 'White' muscle has a high myosin ATP-ase activity and glycogenolytic capacity, but a low hexokinase activity and aerobic capacity. 'Red' fibres show species variation, but tend to the opposite. In 'endurance' training (which for this purpose means repeated bursts of 3–5 min or longer), the main changes are seen in the 'red' fibres. Mitochondrial density and the concentrations of hexokinase, citric acid cycle enzymes, cytochromes and myoglobin increase, whereas glycogen phosphorylase may decrease (see 9.2.1.2). Overall, there is increased storage of glycogen and triglyceride and greater use of free fatty acids as a substrate. Glycogen stores are depleted less rapidly, and less lactate is produced at a given intensity of exercise. On the other hand, higher lactate concentrations (of 20 mmol $\cdot$ litre^{-1} or more) can be tolerated. Thus a greater continuous intensity of exercise can be sustained over longer periods.

An increased aerobic capacity implies an increased supply of oxygen. Part of this results from greater desaturation of muscle venous blood in trained subjects, but the mechanism is not quite clear. There is probably some increase in total capillary density (to be distinguished from opening existing but dormant capillaries (see 9.2.2.2)). The greater mitochondrial density may also enhance diffusion by lowering intramuscular PO_2. There is evidence of a larger shift of the oxyhaemoglobin dissociation curve under working conditions, associated with an increased concentration of 2, 3-diphosphoglycerate in red cells; however, this shift is not apparent under standard conditions *in vitro*. Following training, there is a greater maximal blood flow to muscle, related in part to the greater muscle mass. On the other hand, the blood flow at a given work load is not higher in the trained subject. It may even be lower, because each volume of blood unloads more oxygen.

9.4.2.3 *General changes*. The changes described above (apart from that in the oxyhaemoglobin dissociation curve) are specific to the trained muscles; more general cardiovascular changes may occur. Two or three months of endurance training, involving a large proportion of total muscle mass and near-maximal rates of oxygen consumption, increases both maximal cardiac output and oxygen uptake (see Figure 9.8a). Cardiac output at a given work load is not higher and may be lower. The training should be heavy enough to elicit near-maximal heart rate and oxygen consumption for a few

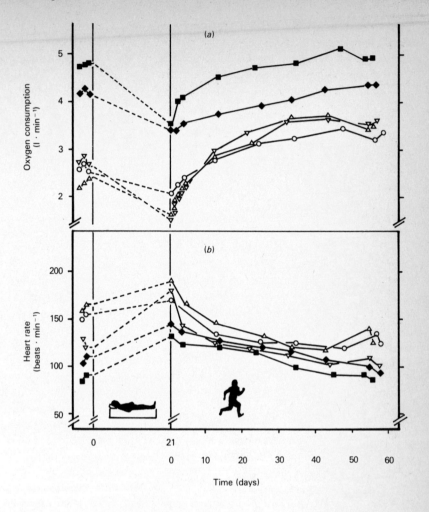

Fig. 9.8 Effects of bed-rest and training on (*a*) oxygen uptake during maximal treadmill exercise and (*b*) heart rate during submaximal exercise (an oxygen consumption of 1.5 litres·min^{-1}). Before the study, three subjects (○, △, ▽) were healthy but athletically inactive; the other two (■, ◆) were highly active. The effects of 3 weeks bed-rest followed by 2 months of strenuous training were qualitatively similar in all subjects, though the changes were relatively greater in the non-athletic. Two subjects (△, ▽) fainted when first exercised after bed-rest. Redrawn from Saltin, B., Blomqvist, G., Mitchell, J.H., Johnson Jr., R.L., Wildenthal, K. & Chapman, C.B. (1968). Response to exercise after bed-rest and after training. A longitudinal study of adaptive changes in oxygen transport and body composition. *Circulation* **38**, No. 5, Supplement No. VII, by permission of the American Heart Association Inc.

minutes, but need not be so heavy or prolonged that the subject is exhausted. Submaximal exercise or training of smaller muscle masses has less effect, presumably because total cardiac output is not a limiting factor. Longer-term training is associated with enlargement of the heart; in endurance exercise the change is in end-diastolic volume, but in static exercise the change is in ventricular wall thickness, reflecting the high arterial pressures that occur (see 9.3.2.4 and Figure 9.4).

Maximal heart rate does not change with training. However, a characteristic effect of training is a lower resting heart rate, which implies a larger stroke volume. As a result, the trained subject can achieve a given work load and cardiac output at a lower heart rate (see Figure 9.8*b*). There is also a proportionate reduction in splanchnic and renal vasoconstriction. Quite light training can lower heart rate: in a group of habitually sedentary women, 5 min of rope skipping each day for a month reduced mean heart rate during the exercise from 168 to 145 beats $\cdot$ min^{-1}. Since the maximal heart rate is unchanged, there is scope for a greater maximal cardiac output and work load. The mechanism of the lower resting heart rate is uncertain: possibly there is a higher vagal tone. It may, however, be just one manifestation of a wider change in autonomic, particularly sympathetic, activity.

In heavy exercise of small untrained muscle groups (e.g. the arms), heart rate and ventilation are higher in relation to work load and oxygen consumption than when large muscle groups are used (see 9.3.2.4). After training of these muscles, the rises in ventilation, heart rate and blood pressure are smaller and there is less splanchnic vasoconstriction. These changes do not 'carry over' to exercise in general, unlike the changes that follow maximal training of large muscle masses.

If training ceases, the heart rate reverts towards its pre-training level in the course of a few weeks (see Figure 9.8). Bed-rest produces a corresponding temporary rise of heart rate, at rest and in exercise. The temporary nature of these changes, and their specificity to particular muscle groups, do not suggest that they are due to permanent structural modifications of the heart or vascular system. Rather, they may represent a 'learning' process which modifies the autonomic and respiratory responses to exercise.

There is often some increase in blood volume and total red cell mass (though not in haematocrit or haemoglobin concentration). Tolerance of dehydration and the capacity to sweat may be greater with long endurance training. There are no changes in lung volumes, or in ventilation or diffusing capacity except those that follow from increased maximal cardiac output and oxygen uptake.

9.4.2.4 *Clinical aspects.* For a patient with cardiorespiratory disorder, everyday tasks may constitute maximal exercise. The burden can be eased by avoiding or modifying some activities. For example, exercise involving small muscle groups or sustained contractions can put greater stress upon the heart than dynamic exercise of large muscle groups (see 9.2.3.4, 9.3.2.4 and Figure 9.4). It may also be less exhausting to work hard intermittently than to work continuously at a nearly maximal load (see 9.3.1.3).

Heart rate is a useful simple criterion of the stress imposed by dynamic exercise. Very approximately, a heart rate 30 beats $\cdot$ min^{-1} below the expected maximum (see 9.2.3.2) corresponds to 70 per cent maximal oxygen consumption. This is usually a safe work load for a few minutes, though a subject with respiratory (rather than cardiovascular) disorder may become intolerably breathless before reaching this level. In any case, it is wise to raise the load in steps and to stop if any distress or cardiac irregularity appears. Particular care is needed where the subject has recently been confined to bed for *any* reason. In five healthy subjects, maximal stroke volume and cardiac output decreased by 5–49 per cent during three weeks of bed-rest, and two of the subjects fainted upon maximal exercise (see Figure 9.8 and Chapters 7 and 8).

9.4.3 *Work and play*
Daily energy expenditure may range from 5 MJ per 24 h (equivalent to an average power of 60 W) for a small female student in a hot climate, to 25 MJ (300 W) for a large Scandinavian lumberjack in winter. More typical would be 8–10 MJ (90–115 W) for women and 10–12 MJ (115–140 W) for men. Within the day there are even wider variations in power consumption. Whereas the student's rate could barely fall any lower (perhaps 50 W, in sleep), the lumberjack skiing to work or wielding his axe may reach 1.3 kW. Suppose he sleeps 8 h (3 MJ) and eats, drinks and plays cards for 8 h (5 MJ). His average power consumption during the 8-h work shift will be only 590 W, i.e. he spends more than half the shift leaning on his axe, getting his breath back. As he tires or ages, he will reduce his average work rate by taking longer rest periods (or by promotion to foreman). Running steadily at 8 km $\cdot$ h^{-1} (5 m.p.h.) uses energy as fast as the lumberjack's average rate, but is less exhausting because the runner is using only 50–60 per cent of maximal aerobic capacity and little lactate is formed (see 9.3.1.3). This illustrates two important points: that the duration of exercise is at least as significant as its intensity in determining the total work done, and that some ways of working are more exhausting than others.

Similar principles apply to play, with the difference that the player may have more latitude in adjusting average power consumption. For example, in a hard game of squash, the rate can be as high as 1.2 kW, but in a desultory game with frequent pauses, it will be much lower. In some team games (e.g. football), the average rate may be quite low, many players using more energy standing shivering than in chasing the ball. On the other hand, the total energy expenditure of a novice playing a full round of golf may be high, due to surplus strokes and searching for lost balls.

Apparently similar sports can make different demands. Cross-country skiing requires a very large aerobic capacity: one skier had a maximal oxygen uptake of 7.4 litres $\cdot$ min^{-1} (equivalent to 2.47 kW). Downhill skiers do not have such large aerobic capacities, but show great isometric strength in the extensor muscles of their legs. Walking is more economical than slow running (or jogging) but the energy cost rises sharply with speed; for example, from 150 J $\cdot$ m^{-1} at 4 km $\cdot$ h^{-1} (2½ m.p.h.) to over 300 J $\cdot$ m^{-1} at 8 km $\cdot$ h^{-1} (5 m.p.h.). The cost of running for the same 75 kg subject was fairly constant at 300 J $\cdot$ m^{-1} over a range 4–14 km $\cdot$ h^{-1} (2½–8½ m.p.h.). Moral: If you get tired running, don't slow to a jog but walk. On the other hand, if you are trying to slim, jogging for long distances is a good way of burning off fat, whereas sprinting would use mainly carbohydrate (see 9.2.1).

In most sports, the fat subject carries a built-in weight penalty; in swimming, this is not so. There are even advantages, in that less energy is used to keep the swimmer afloat and (in British coastal waters) at the right working temperature (see 5.2.4.3). This may be why women are somewhat more efficient swimmers than men. Style and technique make large differences to efficiency. Freestyle (crawl) is more efficient than breaststroke, and butterfly is particularly inefficient. Whereas the skilled runner or cyclist is not much more efficient than the untrained subject, there are large differences between swimmers. In freestyle, the leg stroke is almost irrelevant, unless flippers are worn. With arms alone, the speed is almost as high and the efficiency is higher; the main effect of the leg stroke is to keep the body horizontal. Correspondingly, trained swimmers have a high proportion of 'red' fibres in their shoulder muscles but not in their thighs.

In conclusion, if you choose a sport as a means of keeping fit, decide what you want to achieve first. On the other hand, if you just want to enjoy it, take your pick.

9.5 Further reading

Åstrand, P.-O. & Rodahl, K. (1977). *Textbook of Work Physiology*, 2nd edn. McGraw-Hill: New York.

Cotes, J.E. (1979). Assessment of respiratory control and the physiological response to exercise. In: *Lung Function: Assessment and Application in Medicine*, 4th edn, Chapter 12. Blackwell: Oxford.

Holloszy, J.O. & Booth, F.W. (1976). Biochemical adaptations to endurance exercise in muscle. *Annual Review of Physiology* **38**, 273–91.

Neely, J.R. (1979). Control of carbohydrate and fatty acid metabolism in muscle. In: *Best & Taylor's Physiological Basis of Medical Practice*, 10th edn, ed. J.R. Brobeck. Section 7, Chapter 11. Williams & Wilkins: Baltimore.

Rowell, L.B. (1974). Human cardiovascular adjustments to exercise and thermal stress. *Physiological Reviews* **54**, 75–159.

Robinson, S. (1980). Physiology of muscular exercise. In: *Medical Physiology*, Vol. 2, 14th edn, ed. V.B. Mountcastle. Chapter 58. C.V. Mosby: St. Louis.

Scheuer, J. & Tipton, C.M. (1977). Cardiovascular adaptations to physical training. *Annual Review of Physiology* **39**, 221–51.

Shepherd, R.J. (1978). *Human Physiological Work Capacity*. Cambridge University Press: Cambridge.

Chapter 10

Injury (shock)

Summary

The response to injury has two components, the local response in the injured tissues and the general responses elicited in the rest of the body by the local response. The general response, which is commonly referred to as 'shock', is a complex but co-ordinated pattern of neuroendocrine and autonomic nervous system activity.

The most important aspect of the local response is fluid loss from the circulation, either as blood or as an exudate following an increase in the permeability of the walls of the microcirculation (inflammation). The magnitude of fluid loss can be influenced by many factors including age and temperature but most importantly by the balance of forces in the Starling equilibrium.

Fluid loss from the circulation, which leads to a reduction in blood pressure and volume, is the most important stimulus to the general response to injury. Indeed, many of the physiological changes in the general response are involved in compensating the fluid loss and maintaining an adequate circulation to vital areas such as the brain and heart. The function of the homoeostatic reflexes may be influenced by the presence of toxic factors or by changes in the functioning of the central nervous system.

The metabolic response can also be divided into distinct phases. Injury, or even the anticipation of injury, leads to the mobilization of the body fuels, glucose and fat, as part of the fight or flight response. This is followed by the 'ebb' phase, a period when there is a need to conserve fuels, and this is achieved by a reduction in metabolic rate and the inhibition of glucose utilisation in, for example, skeletal muscle. If the injury is overwhelming, a phase of necrobiosis occurs, characterized by a progressive failure of oxygen transport and death. Recovery involves a period of tissue repair and increased tissue metabolism (the 'flow' phase).

10.1 Introduction

'Fifty per cent of all children born today in the U.K. will be injured or killed in a road traffic accident.'

World Health Organisation (1977)

Accidental injury is the commonest reason for attendance at a hospital casualty department in Western Europe. In the United Kingdom approximately 4 million people are sufficiently injured to attend hospital each year. Of these, 400 000 are admitted and at least 14 000 die. One half of these deaths are a result of road traffic accidents, which are the single largest cause of deaths in males up to 40 years old. In this age group, the fatal injuries are major or multiple; but it is worth remembering that less severe injuries, such as a fractured neck of femur, have a one-year mortality rate of 50 per cent in the elderly (those aged over 65 years). Perhaps the greatest burden that accidents impose on society is the expense of prolonged in-patient treatment and rehabilitation, together with the loss of working time, which is much greater than that lost through industrial disputes.

The responses of the body to injury can be most conveniently divided into *local* (i.e. responses in the injured tissues) and *general* (the response elicited in the rest of the body by the local response). The latter can be referred to as 'shock' and it is in this context that the term has been used in this chapter. Many forms of 'shock' can be distinguished, each of which describes the pattern of responses to a particular injury or insult: for example, traumatic 'shock' (response to trauma), haemorrhagic 'shock' (response to haemorrhage), septic or endotoxic 'shock' (response to infection) and cardiogenic 'shock' (response to heart failure). Attempts to define 'shock' on the basis of a reduced cardiac output (or, more accurately, tissue perfusion) may have more scientific appeal but they are not very satisfactory as they presume knowledge of what is an adequate cardiac output at all stages of the response to injury. There is also the complication of septic 'shock', where cardiac output and tissue perfusion may well be raised. Irreversible 'shock', which refers to the terminal phase of the response preceding death, is characterized by a reduced cardiac output and failure of tissue perfusion (see 10.4.3.4).

The following discussion of some aspects of the response to injury is intended both to stimulate interest in this neglected area of study and to illustrate the challenge that injury provides to our homoeostatic reflexes.

10.2 The local response to injury

When a tissue suffers direct injury, it soon shows the four classic signs of inflammation described in the first century AD by Celsus. They are redness (or flare) and swelling with increase in tissue temperature (heat) and pain. These may be followed by a fifth sign, loss of function, added by Virchow in the mid-nineteenth century. The first three of these signs can be directly attributed to changes in the microcirculation (the venules, capillaries and arterioles) in the injured tissue. The redness is due to vasodilatation of arterioles, heat to increased blood flow, and swelling to an increase in the extravascular fluid content of the injured tissues (oedema). This post-traumatic oedema is a result of an increase in microvascular permeability and is therefore an exudate. Of course, if blood vessels are directly damaged and ruptured, whole blood will also be lost at the site of injury, either escaping to the outside or being retained in the tissue as a bruise or haematoma.

Although the rest of this brief discussion of the local response to injury will be limited to inflammation, it should not be forgotten that fractures of the skeleton and direct damage to underlying tissues such as muscle and deeper organs (e.g. liver and brain) can occur as part of a localized response to injury.

10.2.1 *Vascular changes in inflammation*
The increase in microvascular permeability can be caused by two mechanisms. First, injury may cause direct damage to the endothelium; this can affect all vessels in the microcirculation. Secondly, there is indirect damage which is secondary to the release of an intermediary or mediator at the site of injury. This only affects the venules and is characterized by the pattern of response to classical factors that increase permeability such as histamine, 5-hydroxytryptamine and bradykinin.

Histamine, released from mast cells in the injured tissue, has two main effects on the microcirculation. It causes arteriolar dilatation (hence the redness or flare of the response) and contraction of the endothelial cells in the non-muscular venules. When they contract, the endothelial cells pull away from each other creating 'leaks' in the vessel walls.

What happens after injury will depend on its severity (Figure 10.1). Following a very minor injury, or at the edge of a more severe injury, or after the injection of histamine into the skin, there is an immediate increase in venular permeability. This reaction is transient, lasting only 15 min. However, if the injury is more severe, this early phase is followed by a second, delayed increase in venular and/or capillary permeability which may be prolonged and therefore of greater clinical significance. This second phase is not mediated by factors such as histamine; indeed, no chemical mediator of this delayed phase has been identified. If the injury is very severe, such as a major burn, the immediate and delayed phases become indistinguishable and there is rapid and sustained loss of fluid from all vessels in the microcirculation.

It follows from the above that the use of anti-inflammatory drugs such as antihistamines is of very limited value in the control of post-traumatic fluid loss.

10.2.2 *Modification of the local response*

Fluid loss can be modified by a number of factors. Some of these are included in Starling's equation, as modified by Landis and Pappenheimer, which describes the movement of fluid across the microvascular endothelium as a balance between a number of forces:

$$FM = K (P_c - \pi_{PL} - P_{IF} + \pi_{IF})$$

where FM = fluid movement; K = capillary filtration coefficient (a measure of microvascular permeability and of the surface area available for exchange); P_c = capillary pressure; π_{PL} = plasma protein osmotic pressure; P_{IF} = interstitial fluid pressure (tissue pressure); and π_{IF} = protein osmotic pressure in interstitial fluid immediately outside vessels.

It is difficult to decide what effect the capillary filtration coefficient and osmotic forces will have when selective permeability has been reduced or lost after injury. However, accumulation of protein-rich exudate in the damaged tissue may be a factor tending to move and retain fluid within the extravascular space. The picture is a little clearer when we consider the hydrostatic pressures. If capillary hydrostatic pressure is increased, fluid movement out of the circulation will increase. Conversely, with a low capillary pressure, leakage may not occur. Thus the true extent of tissue injury may be masked in accident victims with severely reduced blood pressure, in whom oedema only becomes obvious when they are resuscitated and blood pressure rises.

A rise in tissue hydrostatic pressure will tend to limit post-traumatic fluid loss from damaged vessels. This mechanism provides a natural limitation to

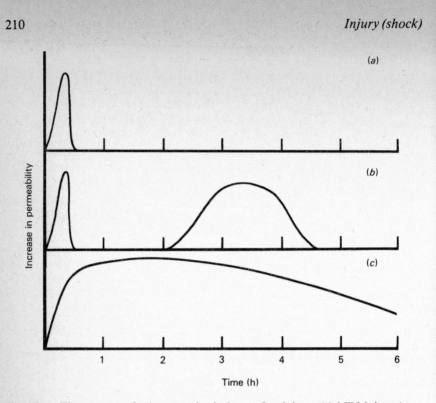

Fig. 10.1   Time course of microvascular leakage after injury. (*a*) Mild injury (e.g. the intradermal injection of histamine). An immediate but transient increase in microvascular permeability confined to the venules; illustrated in photomicrograph (i) by the carbon labelling of leaking vessels. (*b*) Moderate injury (e.g. sunburn). A similar early response to that seen in (*a*) but followed by a later prolonged increase in capillary permeability. (*c*) Severe injury (e.g. major burn). The two phases are no longer distinct and both capillaries and venules are immediately involved; illustrated in photomicrograph (ii) by the carbon labelling of leaking vessels.

the amount of swelling that can occur after injury and can be assisted by the application of external strapping. The rise in tissue pressure and hence the amount of oedema will largely depend on the laxity or compliance of the tissues; for example, compare the marked swelling around the injured eye, where the tissues are lax, with that over the shin, where the tissues are tightly bound together. Oedema will also be influenced by the ability of the lymph vessels to carry away the extra tissue fluid and return it to the circulation through the thoracic duct.

Other factors that can modify the magnitude of fluid loss after injury include tissue temperature, age and the presence of a pre-existing disease. A reduction in temperature of the injured tissue, occurring as the result either of an overall fall in body temperature (hypothermia) or the local application of cold, will reduce oedema formation. The permeability of the micro-

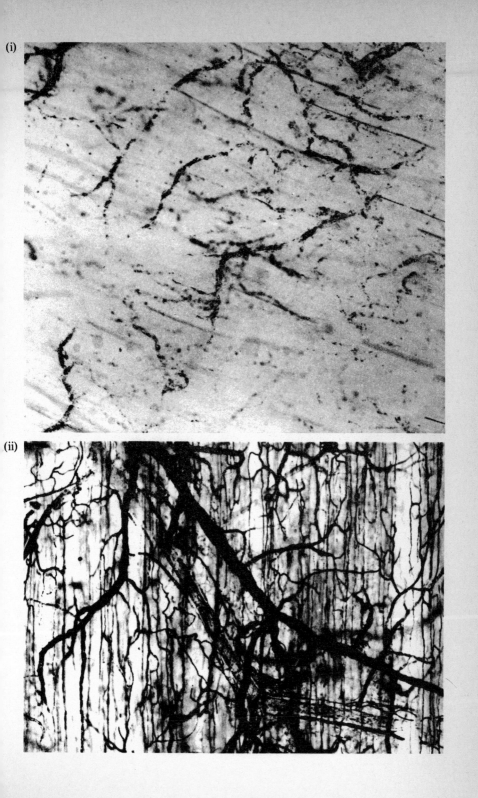

(i)

(ii)

circulation of the newborn and young infant is not increased by factors such as histamine. However, post-traumatic fluid loss is greater in the very young and this may well be due to the greater laxity of the tissues at that age. A disease such as diabetes mellitus modifies the local response to injury; the predisposition of diabetics to infection is a result of their impaired ability to mount an inflammatory response.

10.2.3 *Cellular phase of inflammation*
Although the most obvious external sign of inflammation is swelling, of more importance for defence against the effects of local tissue injury is the accumulation of white blood cells (leukocytes) at the site of injury. The circulating leukocytes first adhere to the blood vessel walls and then migrate across the venular walls between endothelial cells. Once in the injured tissues, the leukocytes phagocytize cellular debris and any invading bacteria.

10.3 The cause of the general response to injury (initiating factors)

What feature or features of the local response are responsible for triggering the general response in the rest of the body? There are three candidates:

(1) fluid loss from the circulation;
(2) release of toxic factors from the injured tissues;
(3) afferent nervous impulses changing the functioning of the central nervous system (CNS).

10.3.1 *Fluid loss from the circulation*
Fluid loss from the circulation after injury can be as whole blood or as a modified form of plasma. The volume of fluid lost after a large burn or when a major vessel is ruptured is obviously considerable. What is not always appreciated is the magnitude of the loss frequently associated with less severe injuries (Table 10.1). For example, after a closed fracture of the thigh bone, the loss can be as high as 2 litres, which is a large amount when

Table 10.1.   Fluid loss after injury in adult man.

Injury	*Fluid loss* (litres)
Fracture-dislocation, ankle	0.25–0.5
Closed fracture, lower leg	0.5 –1.0
Closed fracture, thigh	0.5 –2.0
Open (compound) fracture, thigh	1.0 –3.0
Multiple rib fractures	1.0 –2.0

one considers that the total blood volume of a 70 kg man is only approximately 5 litres.

It does seem that fluid loss from the circulation is the most important cause of the general response to injury. Consequently, a major component of the general response is concerned with the compensation of this fluid loss in an attempt to maintain tissue blood supply; this will be discussed later (see 10.4.1).

10.3.2 *Toxic factors*

The search for toxic factors was largely stimulated by Cannon and Bayliss, who considered that the fluid loss after injury was insufficient to account for the circulatory failure frequently seen in battle casualties during the First World War. If they had measured intravascular volumes, as others did during the Second World War, they would have appreciated the true magnitude of the loss. The concept of a toxic factor being generated in the injured tissues and liberated into the circulation to 'poison' the rest of the body is very plausible. Consequently, many factors have been suggested, ranging from simple ions (e.g. K^+), hormones (histamine, noradrenaline), and metabolites (lactate) to factors extracted from injured muscle or burned skin and bacterial endotoxins. Although many of these candidates will produce 'shock-like' states when given in large doses, they do not do so after doses that are likely to be released or produced by trauma.

Myocardial depressant factor has attracted considerable interest, especially as it has been found in man. It is a low molecular weight peptide or glycopeptide produced in the ischaemic splanchnic area, most probably the pancreas. However, its significance in 'shock' is not clear; the myocardial depressant activity refers to its action in an isolated heart muscle preparation and not necessarily to its activity in vivo. Endogenous opioids may also have a 'toxic' depressant action on the cardiovascular system. The specific opioid antagonist naloxone raises blood pressure and improves survival after some forms of injury. It also raises blood pressure in patients with septic complications following surgery; however, this effect of naloxone is only transient and survival is not improved. In conclusion, it seems unlikely that toxic factors initiate the general response to injury but they may be important later in determining the outcome.

10.3.3 *Afferent nervous impulses and CNS function*

At the end of the last century, Crile concluded that 'surgical shock was due to an impairment or breakdown of the central vasomotor mechanism due to excessive nociceptive afferent stimuli'. More recently it has been demonstrated that the ability to withstand haemorrhage is impaired by

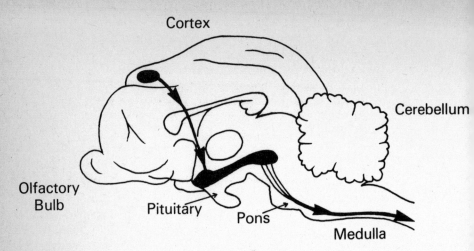

Fig. 10.2 Schematic diagram of corticospinal connections for the hypothalamic defence reaction. Modified from Folkow, B. & Neil, E. (1971). *Circulation*. Oxford University Press: Oxford.

simultaneous nociceptive stimuli. These nociceptive afferent impulses from the injured tissue travel to the brain along the spinothalamic tracts of the spinal cord. Once in the brain these afferents seem to have many targets, including the hind-brain, hypothalamus and even the sensory cortex (Figure 10.2).

If injury has been anticipated, the hypothalamic defence area will have been triggered by impulses from higher centres. The defence reaction, or preparation for fight or flight, involves a mass excitation of the sympathetic nervous system so as to increase blood pressure and cardiac output and redistribute blood from areas such as skin and the splanchnic area to the skeletal muscles, where the blood vessels are dilated by sympathetic cholinergic fibres. It is important to emphasize that there is an increase in both blood pressure and heart rate; the normal baroreceptor-mediated reflex bradycardia in response to a rise in blood pressure is inhibited during the defence or alerting reaction. There is, therefore, a suggestion that normal homoeostatic reflex activity may be modified in response to trauma. Other afferent impulses that may be associated with injury, such as stimulation of a somatic nerve (e.g. sciatic) or of the peripheral chemoreceptors, can also modify the baroreceptor reflex.

Although fluid loss from the circulation is the most important cause of the general response to injury, it is important to be aware of the fact that the ability to compensate for the fluid loss may be modified by changes in the functioning of the central nervous system and by the presence of toxic factors.

10.4 **The general response to injury**

10.4.1 *Compensation of fluid loss*

A loss of fluid from the circulation reduces the venous return of blood to the heart. This reduces cardiac output, which in turn leads to a fall in arterial blood pressure. This lowering of blood pressure is sensed by the high-pressure baroreceptors in the aortic arch and carotid sinuses. The number of afferent impulses from these stretch receptors decreases and so their inhibitory influence on autonomic sympathetic outflow from the central nervous system is reduced. Hence a fall in blood pressure elicits a reflex increase in the activity of the sympathetic nervous system (Figure 10.3). The precise mechanisms involved are exceedingly complex but it seems that the baroreceptor afferents terminate in the nucleus tractus solitarius of the medulla. There is then a diffuse pattern of interconnections within the brainstem, finally converging on the autonomic preganglionic neurones.

The increase in sympathetic activity is enhanced by stimulation of the peripheral chemoreceptors in the carotid and aortic bodies. The chemoreceptors are sensitive to hypoxia, which can be caused either by a reduction in arterial oxygen tension or by a local reduction in blood flow to the chemoreceptors. Although an increase in either hydrogen ion concentration or carbon dioxide tension in the arterial blood also stimulates the peripheral chemoreceptors, their main effects will be mediated by central chemoreceptors located on the ventral surface of the medulla. The important role chemoreceptor activity has in maintaining the increased sympathetic activity is demonstrated by the fall in blood pressure that can occur in the hypo-volaemic experimental animal or man when the oxygen content of the inspired air is increased.

What is the function of this increase in sympathetic activity? It increases the rate and strength of cardiac muscle contractions (i.e. it has positive chronotropic and inotropic effects). By causing peripheral vasoconstriction it also reduces the intravascular space. Constriction of the veins (capacitance vessels) is most important here as some 70 per cent of the blood volume is normally contained within them. There will also be arteriolar constriction, which increases pre-capillary resistance, thereby lowering capillary hydrostatic pressure (the latter determined by the ratio between pre-capillary and post-capillary resistance). From the Starling equation (see 10.2.2) it can be seen that this change will favour the movement of tissue fluid into the circulation ('autotransfusion'). This effect is of most importance in skeletal muscle, which is the largest 'reservoir' of fluid for autotransfusion (Figure 10.4). In a 70 kg man, a fall in capillary hydrostatic pressure of only 4–5 mmHg (0.6 kPa) can lead to an autotransfusion of 200–400 ml within 30 min. It is of interest that those

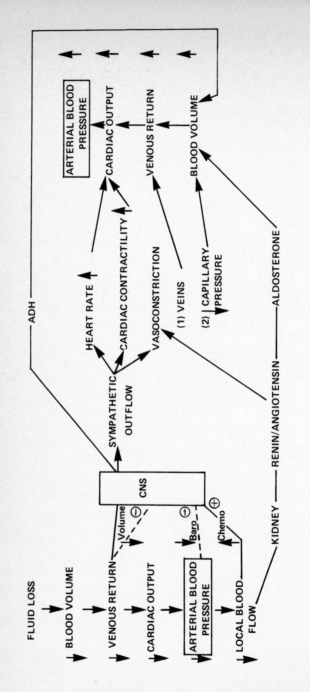

Fig. 10.3   Schematic representation of reflex compensation of fluid loss from the circulation.

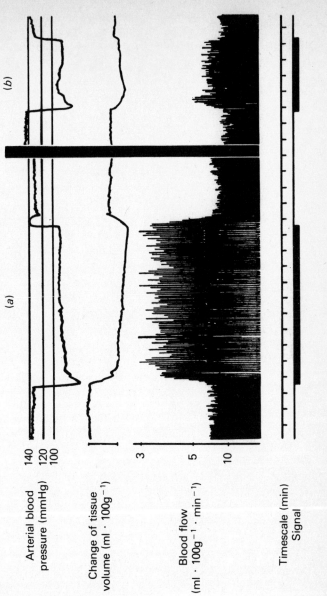

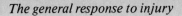

Fig. 10.4 Effect of haemorrhage on the blood pressure and the tissue volume and blood flow in the hind-quarters of the chloralose-anaesthetized cat (*a*) before and (*b*) after cutting the regional sympathetic vaso-constrictor fibres. Note the progressive reduction in tissue volume in the innervated but not in the denervated limbs. From Öberg, B. (1964). *Acta Physiologica Scandinavica* Supplement 229.

species that are most resistant to haemorrhage (e.g. flying birds) are those that can compensate most efficiently for blood loss by rapid autotransfusion. The increased vasoconstriction is not distributed uniformly to all vascular beds: the coronary and cerebral vessels seem to be spared. Thus, in response to a reduction in circulating blood volume, there is a reduction in the size of the vascular bed that has to be perfused; the heart pumps more efficiently, tissue fluid is mobilized and circulation is maintained to vital areas such as heart and brain. If the blood loss cannot be compensated and venous return continues to fall, the heart, under strong sympathetic drive, contracts vigorously but the chambers are not full. This may lead to activation of ventricular distortion receptors which provoke a striking reflex bradycardia (Bezold–Jarisch reflex).

The importance of the sympathetic nervous system in the response to injury is emphasized by the reduced resistance of sympathectomized animals to injury and the poor record of sympathetic blocking agents in the treatment of 'shock'. Unfortunately, vasoconstriction in tissues such as skeletal muscle cannot be maintained indefinitely. Sympathetic stimulation eventually causes a rise in capillary hydrostatic pressure and hence fluid tends to move out of the circulation. This may be analogous to the state of irreversible or decompensated 'shock'. This occurs because the response of the pre-capillary resistance vessels to vasoconstrictor fibre discharge fails before that of the post-capillary vessels. This early failure of the pre-capillary response seems to be caused by the build-up of vasodilator metabolites in the ischaemic tissues. Red blood cells tend to sludge together late in 'shock' and this will tend to raise post- rather than pre-capillary resistance.

Steroid treatment has frequently been advocated in 'shock'. Although this treatment has no sound basis, one beneficial action it may have is the maintenance of vascular responsiveness to sympathetic stimulation.

There are other mechanisms involved in the compensation of fluid loss after injury (Figure 10.3). Renal blood flow is markedly reduced by sympathetic vasoconstrictor fibre activity and the drop in afferent arteriolar pressure leads to an increased output of renin from the juxtaglomerular apparatus. This in turn leads to the formation of angiotensin, a powerful vasoconstrictor agent, which, through the liberation of aldosterone from the adrenal cortex, enhances tubular sodium retention and hence fluid retention. Fluid conservation is also aided by the increase in the secretion of anti-diuretic hormone (ADH) from the posterior lobe of the pituitary gland. One factor involved in the liberation of ADH is stimulation of receptors at the junction between the left atrium and the pulmonary veins (the low-pressure baroreceptors or 'volume' receptors). The injured patient commonly complains of thirst, probably mediated through the 'drinking centres' in the hypothalamus, and will drink avidly. All the mechanisms

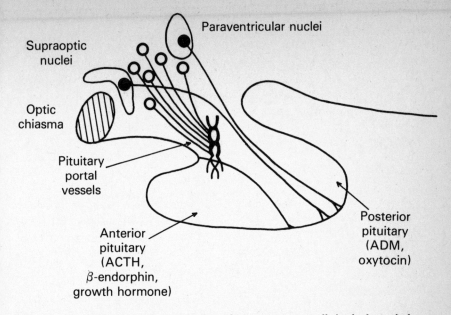

Fig. 10.5 Diagrammatic representation of neurosecretory cells in the hypothalamus showing how the axons of those in the supraoptic and paraventricular nuclei descend to the posterior pituitary whereas the axons of those that secrete hypothalamic-releasing hormones terminate about the vessels of the pituitary portal tract.

described above will help to restore the intravascular volume. Deficits in erythrocytes and plasma protein concentrations require more long-term correction.

10.4.2 *Neuroendocrine responses to injury*

10.4.2.1 *Anterior pituitary gland.* The best known of the endocrine responses to injury are those involving the anterior and posterior lobes of the pituitary gland (Figure 10.5). The secretion of adrenocorticotrophic hormone (ACTH) from the anterior pituitary (adenohypophysis) is controlled by corticotrophin-releasing hormone (CRH) from the hypothalamus; CRH is carried to the pituitary by the pituitary portal vessels. The secretion of CRH is enhanced by injury. This effect seems to be due to afferent stimuli travelling in the spinothalamic tract to the pons and thence, via the medial fore-brain bundle, to the hypothalamus. The neurotransmitters involved in the control of CRH secretion are not all known, although acetylcholine stimulates and both noradrenaline and adrenaline inhibit secretion.

The arterial baroreceptors may provide another link between trauma and ACTH secretion. Stimulation of the baroreceptors due to fluid loss inhibits

ACTH secretion; this may be due to a direct noradrenergic projection from the nucleus tractus solitarius to the hypothalamus.

When ACTH is released from the anterior pituitary, it is accompanied by β-endorphin. As already mentioned (see 10.3.2), the endogenous opioids have been implicated in the development of 'shock' and it is suggested that pituitary endorphins act upon opioid receptors in the brain to depress cardiovascular function.

The secretion of both prolactin and growth hormone is increased in man after injury. However, the relevance of the changes is not known.

10.4.2.2. *Posterior pituitary gland*. The neurones that synthesize ADH and oxytocin and secrete them in the posterior lobe of the pituitary (neuro-hypophysis) have their cell bodies in the magnocellular supraoptic and paraventricular nuclei of the hypothalamus. Both hormones are secreted in response to injury although different stimuli lead to the secretion of ADH and oxytocin in different proportions. The secretion of ADH is normally brought about by stimulation from osmoreceptors and, particularly after fluid loss from the circulation, by afferent vagal fibres from the 'volume' receptors in the thorax. The magnocellular nuclei are stimulated by cholinergic fibres and inhibited by noradrenergic ones.

The blood-pressure-raising and water-conserving actions of ADH are of obvious value after injury. The hormone may also play a role in injury-related metabolic changes such as the conversion of hepatic glycogen to glucose (see 10.4.3.2).

10.4.2.3 *Adrenal gland*. The sympathetic activation that occurs after injury causes an increased secretion of catecholamines from the adrenal medulla. Although both adrenaline and noradrenaline are secreted in response to injury, adrenaline is the principal hormone in the adult. A particularly potent stimulus for secretion is a reduction in blood volume.

As far as the adrenal cortex is concerned, injury stimulates the hypothalamic–hypophyseal system to release ACTH, as outlined earlier, which in turn stimulates the secretion of corticosteroids from the cortex. This adrenocortical response is perhaps the best known response to injury but its function is still not clear. Removal of the adrenal gland (adrenalectomy) greatly reduces resistance to injury. However, adrenalectomized animals maintained on constant doses of cortical hormones show a normal metabolic response to injury. This finding suggests that an increase in the secretion of corticosteroids is not necessary for expression of the metabolic responses to injury and has led to the concept of the 'permissive' role of steroids in 'shock'. Raised concentrations of adrenocortical hormones are, however, necessary for the compensation of post-traumatic fluid loss. Aldosterone is also secreted from the adrenal cortex and, as already mentioned, increased levels of this hormone promote sodium retention in the kidney after injury (10.4.1).

10.4.2.4 *Pancreatic islets*. Injury effects the secretion of both insulin and glucagon from the pancreas. The secretion of these hormones is normally controlled by the plasma glucose concentration. However, it can be modified by the autonomic nervous system. Both stimulation of the sympathetic innervation of the pancreas and circulating adrenaline increase glucagon secretion but decrease that of insulin. Parasympathetic fibres stimulate both insulin and glucagon secretion. Thus, after a mild injury, the plasma insulin concentration may be appropriate for the increased plasma glucose concentration but after more severe injuries, when glucose concentrations are even higher (see below), insulin concentrations may be reduced because of intense sympathetic activation.

10.4.3 *The metabolic response to injury*

10.4.3.1 *Phases of the response*. The first phase of the metabolic response to injury begins with the mobilization of body fuels. In fact mobilization may start before injury as part of the fight-or-flight response discussed earlier. This phase leads into the 'ebb' phase that immediately follows injury and is classically described as a reduction in metabolic rate. Most of the physiological responses already discussed in this chapter occur during this early 'ebb' phase. If the injury is overwhelming, the 'ebb' phase is followed

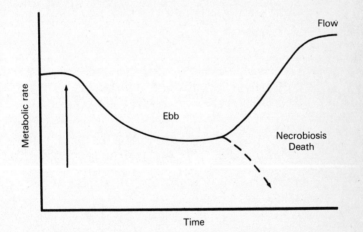

Fig. 10.6 Changes in metabolic rate following injury (arrowed) showing the 'ebb', 'flow' and 'necrobiotic' phases mentioned in the text. The time-course of the phases depends on the severity of the injury and on species. In man, the 'ebb' phase may last for no more than the first 24 h after injury, whereas the 'flow' phase may persist for several weeks.

by a phase of necrobiosis which precedes death. However, with recovery, the 'ebb' phase is followed by the 'flow' phase associated with an increase in metabolic activity (Figure 10.6).

10.4.3.2 *Mobilization of Fuel Stores*. The hydrolysis of tissue glycogen stores to form extracellular glucose and of triglycerides to form free fatty acids is largely due to sympatho-adrenal medullary excitation and the release of hormones.

The principal glycogen stores are in liver and muscle (Figure 10.7). However, these stores are very labile and the increased susceptibility of starved animals to injury may be related to their reduced glycogen stores. The hydrolysis of muscle glycogen after injury is largely due to the action of circulating catecholamines mediated through adrenergic β-receptors. The glycogen is hydrolysed to glucose-6-phosphate, which is converted to lactate and pyruvate (glucose-6-phosphatase is absent in muscle). These can pass into the circulation and are carried to the liver where they are converted to glucose (Cori cycle). The hydrolysis of liver glycogen can be caused by direct stimulation of the sympathetic innervation of the liver and, probably most importantly, by pancreatic glucagon. Anti-diuretic hormone may also be an important factor in stimulating glycogenolysis after injury.

Lipid stores are mobilized after injury by direct activity of the sympathetic innervation of adipose tissue. The triacylglycerol (triglyceride) in adipose tissue is hydrolysed to form free or non-esterified fatty acids and glycerol. The free fatty acids released into the circulation are carried bound to albumin. The rise in plasma free fatty acid concentration after injury will ultimately depend upon the blood flow through the adipose tissue and on the amount of albumin available to bind to the fatty acids. After major injuries and haemorrhage, blood flow in adipose tissue is markedly reduced and the concentration of free fatty acids in plasma does not rise as much as might be predicted from the level of sympatho-adrenal activation.

10.4.3.3 *The 'ebb' phase*. At the time of injury the body's fuel stores may have been mobilized for fight or flight. Since, after injury, it is not possible to replenish the fuel stores, there is a need to conserve or husband any reserves that remain. This is what happens in the 'ebb' phase. The utilization of glucose is inhibited in insulin-sensitive tissues; fat (whose peripheral uptake is not affected by injury) is conserved by a reduction in oxidative metabolism.

In small mammals (e.g. the rat), the 'ebb' phase is characterized by a fall in metabolic rate and body temperature. This is due to a central change in thermoregulation and not to a failure in oxygen delivery to thermogenic tissues. The inhibition of thermoregulation is a result of activation of

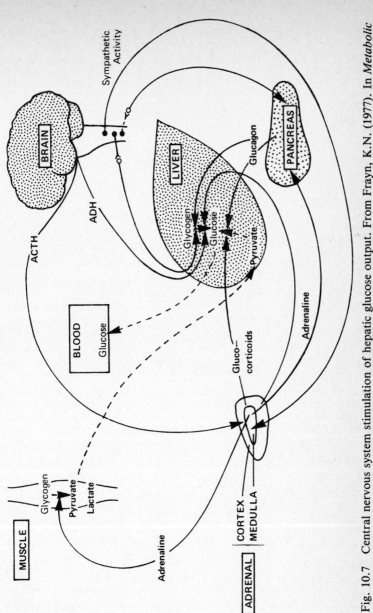

Fig. 10.7   Central nervous system stimulation of hepatic glucose output. From Frayn, K.N. (1977). In *Metabolic Responses to Trauma*. Folia Tramatologia, Geigy: Basle.

noradrenergic fibres supplying the hypothalamus. Although evidence for a similar impairment of thermoregulation in man does not exist, there is some clinical evidence for an inhibition of behavioural thermoregulation after injury. The injured patient selects a higher environmental temperature for comfort than would be predicted from his body temperature. The fall in body temperature seen after severe injuries in man may be due to failure of oxygen transport rather than a central change in thermoregulation. When the temperature of severely injured man does fall, he does not shiver. This may be related to the fact that an adequate arterial baroreceptor input to the central nervous system is necessary for the maintenance of shivering.

The relative contribution of fat oxidation to total energy production is increased after injury. This is reflected in a reduction in the respiratory quotient which occurs very soon after injury. (The respiratory quotient is the volume of CO_2 expired divided by the volume of O_2 consumed: a figure of 1.0 represents carbohydrate oxidation and 0.7 fat oxidation.) The conversion of fatty acids into ketone bodies in the liver is increased after injury. Ketone bodies (aceto-acetate and β-hydroxybutyrate) can form an alternative fuel for skeletal muscle and brain although it is difficult to assess their quantitative importance.

Hyperglycaemia is a prominent feature of the 'ebb' phase. Although adrenergic activity is still high, the main reason for the hyperglycaemia is the reduction in glucose utilization in insulin-dependent tissues such as skeletal muscle. The impairment of glucose utilization is caused by a suppression of insulin secretion by the increased sympathetic activity (an α-adrenergic effect on the insulin-secreting pancreatic β-cells) and to the development of resistance to insulin. This insulin resistance is mediated by the raised levels of growth hormone, adrenaline and the glucocorticoids. The removal of glucose from the blood will also be limited by the increased concentrations of free fatty acids and ketone bodies, whose utilization is limited only by supply (i.e. as plasma concentration rises, tissue uptake also rises). The reduction in glucose utilization by the insulin-dependent tissues will spare this energy substrate for the brain (an insulin-independent organ), which can only use glucose and to a lesser extent ketone bodies.

10.4.3.4 *Necrobiosis*. This is the phase of the metabolic response to injury that immediately precedes death. It is characterized by a failure of oxygen transport to the tissues. Whole-body oxygen consumption and body temperature continue to fall. Increasing hypoxia in the tissues leads to increases in anaerobic glucose utilization, which, with decreased gluconeogenesis, produces the low blood glucose concentrations found terminally. The glucose is only hydrolysed as far as lactate, which accumulates in the circulation. A non-respiratory acidosis develops, with a fall in blood pH and in the total buffering capacity of the body. The lactate : pyruvate and

β-hydroxybutyrate : aceto-acetate ratios rise in blood and tissues, indicating hypoxia in both the cytoplasm and the mitochondria.

10.4.3.5 *The 'flow' phase.* In the 'flow' or recovery phase, metabolic rate may be raised. The fracture of a long bone increases energy expenditure by something like 20 per cent whereas, after a major burn, metabolic rate may be doubled. The increase in metabolic rate may be accompanied by an increase in temperature and raised pulse and respiratory rates. This pattern of response has led to the term 'traumatic fever', which, it must be emphasized, is not related to infection.

During the 'flow' phase, glucose utilization remains impaired and fat is still the most important energy source. However, in order to provide the insulin-independent tissues, such as brain, with glucose, there is a breakdown of muscle protein. This supplies alanine and other amino acids, which are the main precursors for gluconeogenesis in the liver. This mobilization of gluconeogenic amino acids is facilitated by glucocorticoids.

Elevated plasma glucose concentrations normally inhibit the release of the gluconeogenic precursors from skeletal muscle and hence limit hepatic gluconeogenesis. This control mechanism is lost in the injured patient. Thus in the 'flow' phase the rate of fat oxidation and of glucose turnover can both be raised. It has been suggested that this pattern of response is mediated by high plasma glucagon concentrations and especially by the high glucagon : insulin ratio. The increased plasma concentrations of free fatty acids and ketone bodies may help to suppress, or at least limit, skeletal muscle protein breakdown and alanine release.

The loss of muscle protein after major injuries can be large and may even threaten life. The breakdown is reflected in the increased urinary excretion of nitrogen, mainly as urea, but also as creatine, creatinine and 3-methylhistidine. The latter is of interest as it is formed by the methylation of histidine during its incorporation into myofibrillar protein. 3-Methylhistidine is then released when myofibrillar protein is degraded and, as it cannot be reutilized, it is excreted unchanged. It can therefore be used as a quantitative marker of myofibrillar protein breakdown. There are also increased urinary losses of sulphate, phosphate, potassium and zinc.

There is no doubt that patients lose weight after injury and it has been calculated that the loss of lean body mass (predominantly muscle) is almost three times greater than the loss of fat. The loss of body protein is caused by a change in the balance between protein synthesis and protein breakdown. The picture is far from clear but it does seem that, after minor injuries or surgical operations, there is a reduction in muscle protein synthesis, whereas, after more severe injuries, breakdown is also increased. The plasma proteins behave rather differently; some increase after injury whereas others decrease. Those that increase are the acute-phase reactants (e.g. C-reactive protein and fibrinogen) produced in the liver. The acute-phase reactants can be described as the plasma proteins which show an increase in concentration in the acute or early phase of inflammation. Plasma albumin concentration, on the other hand, decreases during the first week or so after injury. This is due both to a reduction in synthesis and to an increase in the movement of albumin into the extravascular space. A link between the inflammatory reaction (where this chapter began) and protein synthesis is provided by leukocyte endogenous mediator released from leukocytes, which stimulates the hepatic uptake of plasma amino acids utilized in the synthesis of acute-phase globulins.

10.5 Further reading

Fine, J. (1966). Shock and peripheral circulatory insufficiency. In: *Handbook of Physiology*, Section 2, Circulation, Vol. III, pp. 2037–69.

Folkow, B. & Neil, E. (1971). *Circulation*. Oxford University Press: London.

Kelman, G.R. (1977). Haemorrhage and shock. In: *Applied Cardiovascular Physiology*, Chapter 9. Butterworth: London.

Landis, E.M. & Pappenheimer, J.R. (1966). Exchange of substances through the capillary walls. In: *Handbook of Physiology*, Section 2, Circulation, Vol. II, pp. 961–1034.

Metabolic Responses to Trauma (1977). Folio Traumatologia, Geigy: Basle.

Spector, W.G. & Willoughby, D. (1968). *The Pharmacology of Inflammation*. English Universities Press: London.

Wiggers, C.J. (1950). *Physiology of Shock*. The Commonwealth Fund: New York.

Index